Physical Chemistry and Acid-Base Properties of Surfaces

why how
do
they
civilizations
of Satan
~~fall~~ like
Anon, Satan's,
fall after to
~~long success~~
bad free choice
after long success -

FOCUS SERIES

Series Editor Michel Pons

Physical Chemistry and Acid-Base Properties of Surfaces

Jean-Charles Joud
Marie-Geneviève Barthés-Labrousse

iSTE

WILEY

First published 2015 in Great Britain and the United States by ISTE Ltd and John Wiley & Sons, Inc.

ISTE Ltd
27-37 St George's Road
London SW19 4EU
UK

www.iste.co.uk

John Wiley & Sons, Inc.
111 River Street
Hoboken, NJ 07030
USA

www.wiley.com

Library of Congress Control Number: 2015952414

British Library Cataloguing-in-Publication Data
A CIP record for this book is available from the British Library
ISSN 2051-2481 (Print)
ISSN 2051-249X (Online)
ISBN 978-1-84821-843-7

Contents

Introduction

Capillarity is the study of interfaces between two immiscible liquids, between a liquid and air or between a solid and a liquid. These interfaces are able to change shape or even move in order to minimize the energy of the system.

This science started to be developed in the 19th Century by its founding fathers: Laplace and Young. More recently, different schools have developed the domain further. These include, among others, the Russian school with Derjagin, the Belgian school with Defay and Prigogine, the French school with Lippmann, Bouasse and de Gennes, and the American school with Fowkes, van Oss-Good and Israelachvili.

The last two decades have witnessed the significant development of studies of wettability, both experimentally and theoretically due to applications in different technological domains. This includes the areas of adhesive bonding in the production of metal/polymer interfaces, microelectronics with the development of molecular bonding, fuel cells with the need to manage water in a porous medium, the development of a knowledge-based approach to the cleanability of materials such as glass or stainless steel, and the value of electrowetting in the development of "Lab-on-a-chip" applications. The field now covered is considerable.

This book provides an overview of a Master's level course developed by one of the authors over a 15-year period with engineering students and PhD students from the Phelma school of the *Institut Polytechnique* in Grenoble, France. The examples given are subjectively chosen by the authors, although they are guided by the importance and originality of the underlying applications.

Several domains are omitted such as those regarding, for instance, the spreading kinetics of a liquid or the elasticity of triple lines. Likewise experimental aspects linked to determining surface tensions are not presented in detail and refer to very thorough existing works which are indicated in the bibliography.

Following a reminder of the fundamental aspects of wettability regarding ideal and real surfaces, we focus on the study of various components of the interfacial tension. A review of the literature and its evolution, through the analysis of quantum mechanics, provides an understanding of Fowkes formulations on the one hand, and Van Oss and Good's on the other hand.

The use of this formalism from an experimental point of view is presented by conducting wettability experiments (one liquid and two liquid methods) and illustrated by examples.

Various experimental methods are then described in order to evaluate the acid-base properties of solid surfaces, on a macroscopic scale (mean overall measure) or on an atomic or molecular scale (identification of the acid or base nature of the various sites present on the surface).

Finally, the problem of charges at the solid–liquid interface, which is rarely addressed in the literature, is examined in detail using the concepts of the electrochemical double layer. This key issue is central to the concept of zero point of charge, i.e. the pH of the aqueous solution in contact which leads to an overall zero surface charge.

The situation of the zero point of charge is also characterized by a maximum in the solid–liquid interfacial tension.

Using these results, the wettability measurements taken as a function of the pH of the aqueous solution help to define the zero point of charge of different surfaces (usually oxides). Examples of model and industrial surfaces are given.

The same concepts also demonstrate how the injection of charges by a potential will lead to a reduction in the solid–liquid interfacial tension thus leading to a decrease in the contact angle at the origin of the electrowetting.

The latter effect is at the source of significant recent developments in the field of microfluidics. Generic configurations used are presented and illustrated by recent works taken from the literature.

1

Wettability of an Ideal Surface: Overview

1.1. Wetting angle

Young's relationship and definition of the work of adhesion.

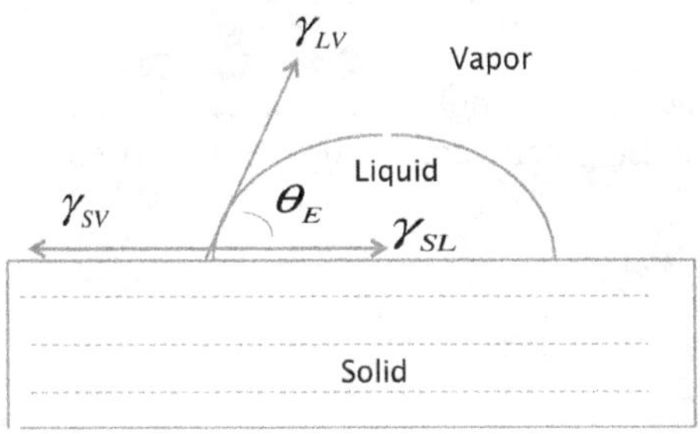

Figure 1.1. *Equilibrium of a liquid droplet on an ideal solid substrate*

When a liquid droplet is placed in a solid, it takes on a given configuration according to the interactions which take place between the solid and the liquid.

The system evolves toward a configuration of minimum free energy characterized by an angle of equilibrium θ_E.

If there is no chemical reaction at the solid–liquid interface, the geometry is simple, as shown in Figure 1.1.

Young's relationship [YOU 05] provides the equilibrium of forces at the triple line. Interfacial tensions are described as forces per unit of triple line length [DEF 66, DE 03]

$$\gamma_{SV} = \gamma_{SL} + \gamma_{LV} \cos\theta_E \qquad [1.1]$$

where θ_E is the wetting angle and γ is the interfacial tension of solid–vapor, liquid–vapor and solid–liquid, respectively.

COMMENT 1.1.– γ_{SV} is the interfacial tension of the solid in equilibrium with the vapor phase. This value is generally different from the surface tension of the solid in equilibrium with vacuum or a neutral gas, denoted as γ_S. This difference is known as spreading pressure: π_e

$$\pi_e = \gamma_S - \gamma_{SV} \qquad [1.2]$$

Young's equation can therefore be written as follows:

$$\gamma_S = \gamma_{SL} + \gamma_{LV} \cos\theta_E + \pi_e \qquad [1.3]$$

The same result as Young's equation can be deduced from a virtual thermodynamic cycle (there are a number of different variations of this demonstration):

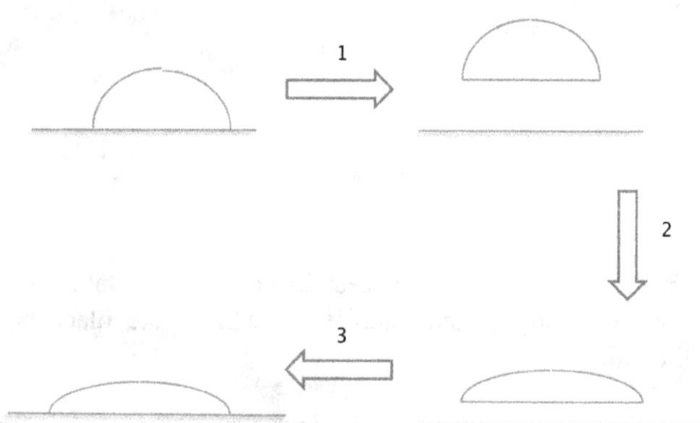

Figure 1.2. *Virtual thermodynamic cycle for the equilibrium of a liquid droplet*

Using a droplet with the interfacial area Ω placed on the solid, it can be assumed first that the droplet can be separated from the substrate without changing its shape.

$$\Delta G_1 = -\gamma_{SL}\Omega + \gamma_{SV}\Omega + \gamma_{LV}\Omega$$

Subsequently, the raised droplet undergoes a variation in contact area dΩ corresponding to the equilibrium situation. As a result, there is a variation in the area of the droplet cap: dΩ cosθ (Figure 1.3).

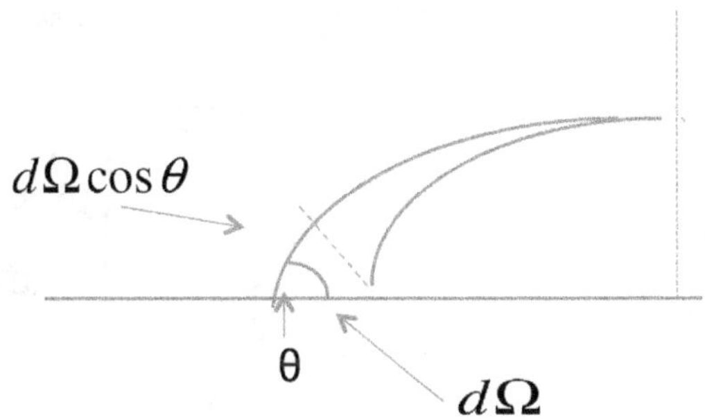

Figure 1.3. *Area variation of the liquid droplet when it reaches the state of equilibrium*

Hence: $\Delta G_2 = \gamma_{LV} d\Omega + \gamma_{LV} d\Omega \cos\theta$.

In the third stage, the droplet is placed back onto the solid:

$$\Delta G_3 = (\gamma_{SL} - \gamma_{SV} - \gamma_{SL})(\Omega + d\Omega)$$

The overall outcome is: $\Delta G = \Delta G_1 + \Delta G_2 + \Delta G_3$.

Hence: $\Delta G = (\gamma_{SL} - \gamma_{SV} + \gamma_{LV} \cos\theta_E) d\Omega$.

At equilibrium, this gives: $d\Delta G / d\Omega = O$.

Hence: $\gamma_{SL} - \gamma_{SV} + \gamma_{LV} \cos\theta_E = 0$.

This does indeed bring us back to Young's equation.

1.2. Adhesion effect

Dupré definition, spreading coefficient.

The work of adhesion W_{SL} is the variation of free energy per unit area involved in the following transformation shown in the diagram in Figure 1.4.

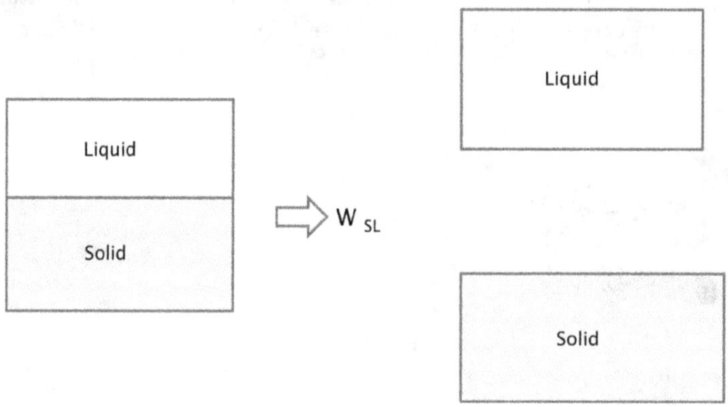

Figure 1.4. *Definition of work of adhesion*

The Dupré relationship is written as follows [DEF 66]:

$$W_{SL} = \gamma_S + \gamma_{LV} - \gamma_{SL} \quad [1.4]$$

In this definition, γ_S corresponds to a flat surface without vapor adsorption. The solid and liquid phases are very far away from each other.

The introduction of Young's relationship thus leads to the following expression:

$$W_{SL} = \gamma_{LV}\left(1+\cos\theta_E\right)+\pi_e$$

This expression is too often reduced to the following one which ignores the spreading pressure without truly considering its relevance (see Chapter 3):

$$W_{SL} = \gamma_{LV}\left(1+\cos\theta_E\right) \quad [1.5]$$

Indeed, the difference $\pi_e = \gamma_S - \gamma_{SV}$ can only be ignored in the case of a solid with low surface energy or if the vapor (water) adsorption only has a minor influence: polymers or oxides with a largely contaminated surface.

The ability of a solid for wetting by a liquid is shown by a strong adhesion effect and a low wetting angle.

Another parameter is also used to calculate wetting, which is the spreading coefficient S_e popularized by de Gennes [DE 03]:

$$S_e = \gamma_S - \gamma_{LV} - \gamma_{SL} \quad [1.6]$$

Using previous relationships, the spreading coefficient can be written as follows:

$$S_e = \pi_e - \gamma_{LV}\left(1 - \cos\theta_E\right) = W_{SL} - 2\gamma_{LV}$$

Extensive spreading of the liquid over the solid is obtained for positive spreading coefficients.

COMMENT 1.2.– Free energy of adhesion ΔF_a which is equal, apart from the sign, to the adhesion effect is a concept which is also widely used in the scientific literature.

1.3. Surface tension and free surface energy

Simplified analysis of surface tension and stresses for a solid or liquid surface. Relation with specific free surface energy.

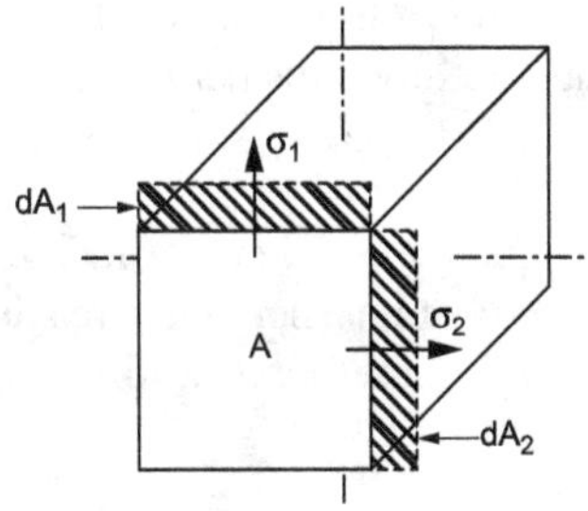

Figure 1.5. *Reversible deformation of a face of a solid crystal*

For the solid surface of a crystal, Figure 1.5 shows the reversible deformation at a constant temperature of the face of the crystal, slightly deformed by quantities dA_1 and dA_2 due to surface stresses σ_1 and σ_2 and in the absence of volume deformation:

$$\sigma_1 dA_1 + \sigma_2 dA_2 = d\left(A.F_A\right)$$

where F_A is the specific free energy of the surface

with $F_A = U_A - TS_A$ (U_A: specific energy of the surface and S_A specific entropy of the surface).

For a symmetrical crystal or an isotropic solid, Shuttleworth shows that [SHU 50]:

$$\sigma_1 = \sigma_2 = \gamma_A$$

$$\gamma_A(dA_1 + dA_2) = \gamma_A dA = d(AF_A)$$

Hence:

$$\gamma_A = F_A + A\left(\frac{dF_A}{dA}\right) \qquad [1.7]$$

For a liquid substrate which undergoes isothermal deformation at a constant volume, the second term of the previous equation is zero. Indeed, the atoms of the liquid are transferred to the surface and the number of atoms per unit area remains constant.

At low temperatures in the case of a solid, atoms are relatively immobile. After deformation, the number of atoms per unit area at a surface layer varies and dF_A/dA is not zero.

However, at high temperatures where diffusion is sufficient, it can be assumed that the number of atoms per unit area in the surface layer will remain relatively constant and therefore $dF_A/dA \approx 0$.

Hence:

$$\gamma_A \approx F_A \qquad [1.8]$$

The complete thermodynamic analysis as proposed by Defay and Prigogine [DEF 66] helps to rigorously establish the useful relationships between the characteristic quantities of the surface.

We refer back to these simply by using their analogy with the thermodynamic relationships of the bulk phase, which are supposedly more familiar.

The definition of free enthalpy G is written as:

$G = F + PV = \Sigma_i n_i \mu_i$ or by volume unit: $G_V = F_V + P = \Sigma_i c_i \mu_i$

F_v represents the free energy per volume unit, c_i is the bulk concentration of the element i and μ_i its chemical potential.

This relation is written as: $-P = F_V - \Sigma_i c_i \mu_i$

By analogy, we obtain the following for the surface phase:

$$\gamma = F_A - \Sigma_i \Gamma_i \mu_i \qquad [1.9]$$

Gibbs adsorptions are given by the expression $\Gamma_i = n_i^A / \Omega$ where n_i^A represents the excess in the number of moles of the constituent I on the uppermost surface compared to the bulk composition $n_i^A = n_i^s - n_i^b$ (rigorously for a division surface placed above the surface layer covering the homogeneous bulk phase) and Ω is the area of the surface.

The surface tension γ, similar, apart from the sign, to the bulk pressure, is expressed according to the specific free energy of the surface F_A, chemical potentials and Gibbs adsorptions Γ_i.

This relation shows that the surface tension γ is only reduced to the specific free energy of the surface F_A for a pure body or systems for which the adsorption Γ_i can be cancelled out by a suitable choice of the position of the division surface.

For multi-component systems where the temperature is such that diffusion can take place, there may be a segregation of one or more surfactant agents toward the surface or interface, causing an enrichment of the interface in surfactant elements and a reduction in the surface tension [DEF 66].

The variation in the free enthalpy G of a bulk phase can generally be written as follows:

$$dG = VdP - SdT + \Sigma_i \mu_i dn_i = d(\Sigma_i n_i \mu_i)$$

After simplification, this relation leads to the generalized Gibbs Duhem relationship, hence per volume unit:

$$dP_V = S_V dT + \Sigma_i c_i d\mu_i$$

By analogy, we can obtain the Gibbs adsorption equation:

$$d\gamma = -S_A dT - \Sigma_i \Gamma_i d\mu_i \quad [1.10]$$

This relationship will be used during the study of the solid–electrolyte interface in Chapter 7.

To conclude this short introduction, we predict that wettability studies will be important in various domains of application in material science.

At high temperatures, studies of this type will be necessary to address the issue of adhesion of different assemblies, such as, for example, ceramic–metal solder joints or the manufacturing of composite materials such as Al/SiC or Al/C.

At lower temperatures (close to ambient temperature), wettability studies help to establish the energy of solid surfaces, which can show structural defects and/or can be coated with a contamination layer (real case). It would therefore be possible to characterize the acid-base properties of the material studied, which is a quantity which guides in particular the adhesion of the now widely-used metal-polymer assemblies.

2

Real Surfaces

The surfaces discussed in the previous chapter are ideal surfaces, meaning that they are atomically smooth and chemically homogeneous.

In practice, solid surfaces are both rough and chemically heterogeneous. These surface defects cause a dispersion of the contact angle around an average value which can be up to ten degrees or even higher. This dispersion, which corresponds to the hysteresis of the contact angle, shows that a single measurement of the angle is inadequate. A significant measurement of the contact angle corresponds to an average value taken from around 10 measurements at different points of the surface being studied, leading to a roughly Gaussian distribution.

There are two models which separately deal with the existence of topological or chemical defects. These models (Wenzel [WEN 49] and Cassie–Baxter [CAS 44]) are exhaustively examined by de Gennes in his 2002 book [DE 03]. We will address them briefly.

2.1. Wenzel's model – topological defects

Apparent contact angle and topological roughness.

The apparent contact angle θ^* is observed on a chemically homogeneous surface with a low level of roughness r compared to the size of the droplet. The roughness r is defined by the ratio between the area of the real surface and that of the projected surface.

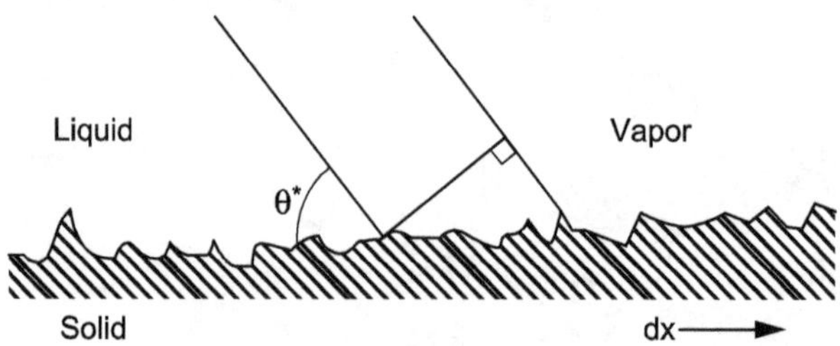

Figure 2.1. *Equilibrium shift of the triple line of a liquid droplet in contact with a rough surface*

The variation in the free energy dF associated with an equilibrium shift dx of the triple line is written per unit of length of the triple line:

$$dF = r(\gamma_{SL} - \gamma_{SV})dx + \gamma_{LV} dx \cos\theta^* = 0$$

since Young's equilibrium contact angle is given by the equation $\gamma_{SV} - \gamma_{SL} = \gamma_{LV} \cos\theta_E$.

We obtain Wenzel's equation [2.1] connecting the apparent contact angle θ^* and Young's angle θ_E :

$$\cos\theta^* = r\cos\theta_E \qquad [2.1]$$

In the case of a hydrophilic solid where $\theta_E \leq 90°$, the apparent contact angle is such that $\theta^* \leq \theta_E$.

Similarly, for a hydrophobic solid characterized by $\theta_E \geq 90°$, the apparent contact angle will be such that $\theta^* \geq \theta_E$. *The roughness of a surface accentuates its specific hydrophilic or hydrophobic nature.*

This simple result is obtained for roughness values *r* clearly higher than 1 and with an upper limit around 1.8-2 as we will see in the following section.

For higher values, an air pocket regime is established leading to the trapping of air pockets by the wetting liquid at the bottom of the cavities.

2.2. Cassie–Baxter model: chemical defects

Apparent contact angle and chemical heterogeneity.

Here, the surface is considered as being atomically flat with supposedly small levels of chemical heterogeneity compared to the size of the liquid droplet.

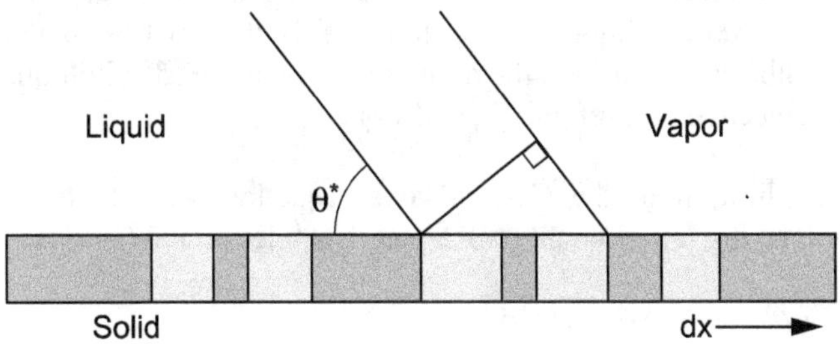

Figure 2.2. *Equilibrium shift of the triple line on a chemically heterogeneous surface*

Let us assume two types of chemical defects with the fractions f_1 and f_2, characterized by Young's angles θ_1 and θ_2. The variation in the free energy corresponding to the equilibrium shift dx is written as follows:

$$dF = f_1(\gamma_{SL} - \gamma_{SV})_1 dx + f_2(\gamma_{SL} - \gamma_{SV})_2 dx + \gamma_{LV} \cos\theta^* dx = 0$$

Using Young's equation for each of the solids 1 and 2, we can obtain the Cassie–Baxter equation [2.2]:

$$\cos\theta^* = f_1 \cos\theta_1 + f_2 \cos\theta_2 \qquad [2.2]$$

This equation shows that the cosine of the apparent contact angle θ^* is given by an average of the cosines of Young's angles θ_1 and θ_2 characterizing phases 1 and 2.

This equation can be very useful for exploring situations corresponding to superhydrophobic and superhydrophilic surfaces.

2.3. Superhydrophilic surfaces

Superhydrophilic rough surface, development of a precursor film.

Definition of a critical angle between a Wenzel state and a Cassie–Baxter state with water pockets.

Let us consider a superhydrophilic porous surface (e.g. a mixture of oxides). The wetting liquid spontaneously fills the cavities of the surface. When equilibrium is reached, the droplet rests on a mixed solid/liquid surface with the respective proportions φ_S and $1-\varphi_S$.

The application of the Cassie–Baxter equation leads to the following expression taking into account both Young's angles θ_E and 0:

$$\cos\theta^* = \phi_S \cos\theta_E + (1-\phi_S) \qquad [2.3]$$

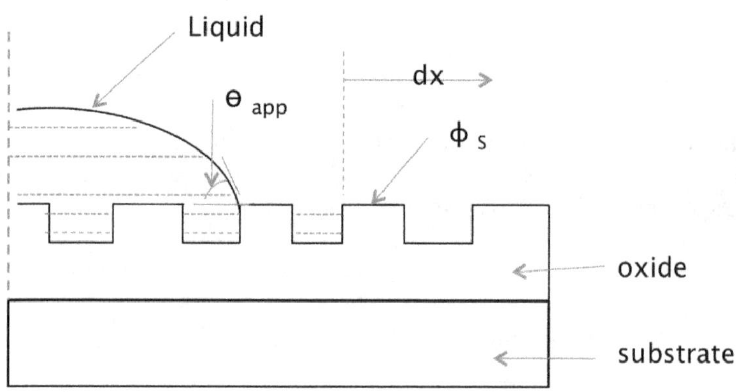

Figure 2.3. *Model of a superhydrophilic rough surface with the development of a precursor film*

The total apparent wetting ($\theta^* = 0$) can only be obtained with the condition $\theta_E = 0$.

The condition for this description to be valid is that a precursor film develops in front of the base of the droplet as indicated by de Gennes [DE 03] and Bormachenko [BOR 08].

In this zone, the vapor pressure of the gaseous phase is given by the equilibrium pressure of the liquid phase. For sufficiently-sized surface pores, the development of a horizontal liquid surface can be observed as indicated in Figure 2.3.

The variation in the free energy (per unit length for the triple line) corresponding to an equilibrium shift dx (Figure 2.3) is written by using r to represent the roughness associated with the solid surface simply modeled by regular cavities.

$$dF = (\gamma_{SL} - \gamma_{SV})(r - \phi_S)dx + \gamma_{LV}(1 - \phi_S)dx \qquad [2.4]$$

The first term corresponds to the creation of a solid–liquid interface on the internal part of the surface, corrected by the fraction $\phi_S dx$ which corresponds to the external dry part of the surface.

The second part corresponds to the creation of a liquid–vapor interface.

Using Young's equation, the stability condition of the pre-wetting film ($dF \leq 0$) gives the following condition:

$$\cos\theta_E \geq \frac{1 - \phi_S}{r - \phi_S} = \cos\theta^0_{crit} \quad \text{et } \theta_E \leq \theta^0_{crit} \qquad [2.5]$$

The critical contact angle θ^0_{crit} evolves between two limits $\theta^0_{crit} = 0$ for a perfectly flat surface (r=1) and $\theta^0_{crit} = \pi/2$ which characterizes a completely porous solid with a high value of r.

Generally, when the Young's angle is between θ^0_{crit} and $\pi/2$, the solid is dry in front of the base of the droplet and the wetting is described by Wenzel's equation. When $\theta_E \leq \theta^0_{crit}$, a pre-wetting film develops in form [2.3] of the Cassie–Baxter equation which describes the wetting of the liquid droplet resting on a liquid–solid composite.

Figure 2.4 shows this evolution.

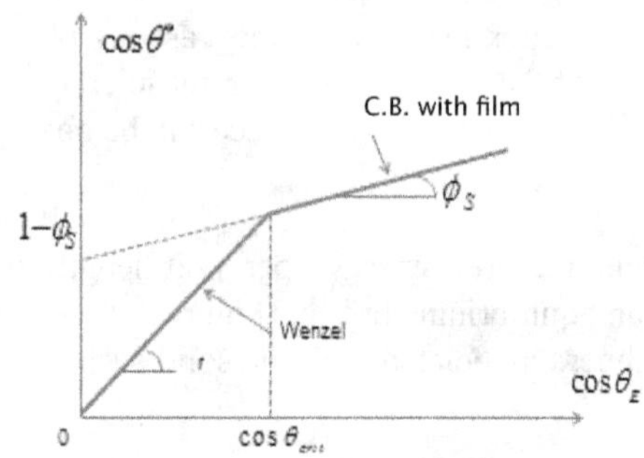

Figure 2.4. *Evolution of the apparent contact angle with Young's angle for a hydrophilic solid*

COMMENT 2.1.– The simplified model of the surface which we have used in Figure 2.3 enables a simple calculation of the critical angle.

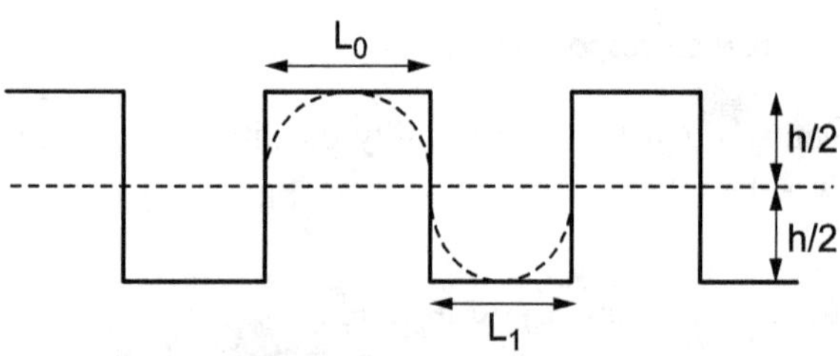

Figure 2.5. *Simplified model of the roughness of the surface*

Using the lengths indicated in Figure 2.5, the roughness coefficient r is written as:

$$r = \frac{L_0 + L_1 + 2h}{L_0 + L_1} = 1 + \frac{2h}{L_0 + L_1}, \text{ with a surface fraction } \phi_S = L_0 / L_0 + L_1.$$

$$\text{Hence: } \cos\theta^0_{cri} = \frac{1-\phi_S}{r-\phi_S} = \frac{1}{1+2h/L_1} \qquad [2.6]$$

Table 2.1 provides an example of some values of critical angles.

COMMENT 2.2.– The presence of a pre-wetting film in front of the base of the droplet leads to a relatively flat liquid–vapor interface. It may develop with the liquid–solid interface at the bottom of cavities long distance interactions, such as Van de Waals and electrostatic interactions denoted as G_{LD} per unit area, which will be described in Chapter 8.

In these conditions, equation [2.5] must be corrected [JOU 13]:

$$dF = (\gamma_{SL} - \gamma_{SV})(r - \phi_S)dx + \gamma_{LV}(1 - \phi_S)dx + (1 - \phi_S)dxG_{LD}$$

The stability conditions for the pre-wetting film $dF \leq 0$ leads to a new condition:

$$\cos\theta_E \geq \frac{1-\phi_S}{r-\phi_S}\left(1+\frac{G_{LD}}{\gamma_{LV}}\right) = \cos\theta_{crit} \qquad [2.7]$$

A new critical angle is thus defined as:

$$\cos\theta_{crit} = \cos\theta^0_{crit} \cdot \left(1+\frac{G_{LD}}{\gamma_{LV}}\right) \text{ avec } \theta_E \leq \theta_{crit} \angle \theta^0_{crit} \qquad [2.8]$$

$G_{LD} / \gamma_{LV} = 0.6$

$2h/L_1$	$\cos\theta^0_{crit}$	θ^0_{crit} deg	$\cos\theta_{crit}$	θ_{crit} deg
0.8	0.555	56	0.88	27
0.9	0.526	58	0.84	33
1	0.500	60	0.80	37
1.8	0.357	69	0.57	55

$G_{LD} / \gamma_{LV} = 0.7$

$2h/L_1$	$\cos\theta^0_{crit}$	θ^0_{crit} deg	$\cos\theta_{crit}$	θ_{crit} deg
0.8	0.555	56	0.943	19
0.9	0.526	58	0.894	26
1	0.500	60	0.850	32
1.8	0.357	69	0.607	53

$G_{LD} / \gamma_{LV} = 0.9$

2h/L_1	$\cos\theta^0_{crit}$	θ^0_{crit} deg	$\cos\theta_{crit}$	θ_{crit} deg
0.8	0.555	56	1.05?	?
0.9	0.526	58	0.999	2
1	0.500	60	0.95	18
1.8	0.357	69	0.68	47

Table 2.1. *Values of critical angles deduced from equations [2.6] and [2.8] for different geometries (Figure 2.5) set according to the G_{LD} / γ_{LV} ratio*

The inclusion of long distance interactions G_{LD} which are comparable in terms of magnitude with γ_{LV} leads to a significant reduction in the critical angle θ_{crit}, thus strengthening the condition $\theta_E \leq \theta_{crit} \angle \theta^0_{crit}$, which must be fulfilled for superhydrophilic behavior.

In these conditions, the apparent contact angle is given by equation [2.3].

$$\cos\theta^* = \phi_S \cos\theta_E + (1 - \phi_S)$$

When Young's angle is low, the previous equation can be drastically simplified (limited development):

$$\text{Hence: } \theta^* \approx \theta_E \sqrt{\phi_S} \qquad [2.9]$$

2.4. Superhydrophobic surfaces

Superhydrophobic surface with air pockets. Roughness boundary between a Wenzel state and a Cassie–Baxter state with air pockets.

For a hydrophobic solid, $\theta_E > \pi/2$, the solid–liquid tension is greater than the solid–vapor tension. The solid/liquid interface must be minimized and may not follow the topography of the solid surface as shown in Wenzel's equation. In other words, the system may find it beneficial to trap air pockets in cavities induced by the roughness, i.e. to develop a small liquid–vapor interface rather than a complete solid–liquid interface.

Figure 2.6 shows two models of the surface in such a situation.

Figure 2.6(a) shows the simplest situation, Figure 2.6(b) shows a more realistic sinusoidal interface.

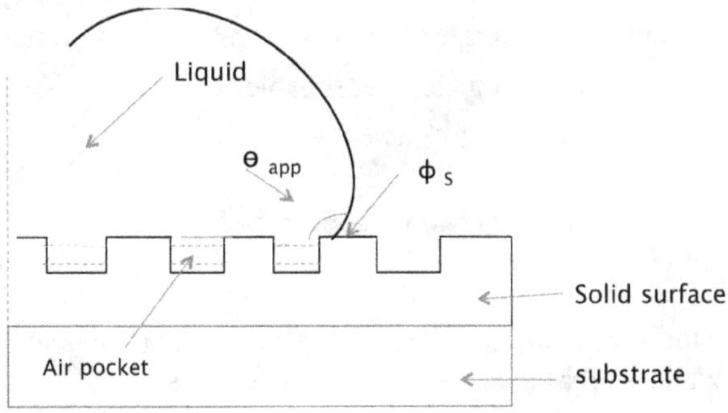

Figure 2.6. *a) Simple hydrophobic surface with air pockets*

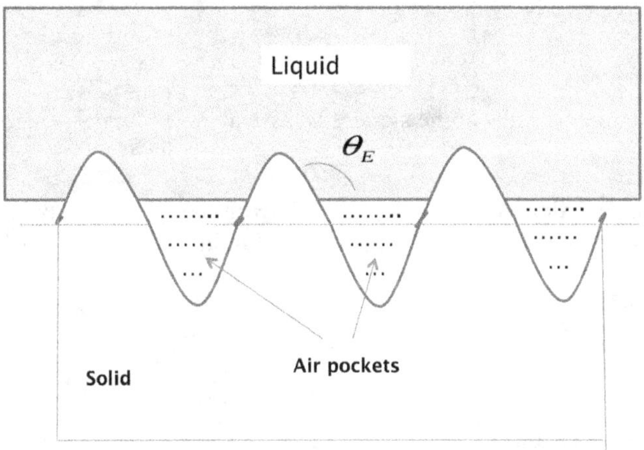

Figure 2.6. *b) Detail of a more realistic roughness*

In the case of the simple model (Figure 2.6(a)) the liquid droplet rests on a solid–air composite. The application of the Cassie–Baxter equation is written in this situation where the two equilibrium angles are θ_E and π:

$$\cos\theta^* = \phi_S \cos\theta_E - (1-\phi_S) \qquad [2.10]$$

For a value of θ_E =120° ($\cos\theta_E = -1/2$), which is a common value for a hydrophobic solid, a simple calculation using equation [2.10] leads to:

$$\theta^* = \pi - \sqrt{\phi_S}$$

hence an apparent contact angle θ^* higher than θ_E in the range 123°–180° depending on the value of ϕ_S. A reasonable value ϕ_S=15% leads to an apparent contact angle θ^*=158°.

The apparent contact angle can therefore be significantly greater than the equilibrium angle.

Using a more realistic solid surface (Figure 2.6(b)) helps to define a threshold roughness for the appearance of air pockets.

For a solid surface given by equation [2.11]:

$$z = a\cos\frac{2\pi}{\lambda}x \qquad [2.11]$$

the maximum modulus of the slope is written as $\left|\frac{dz}{dx}\right|_{\max} = a\frac{2\pi}{\lambda}$

The actual slope must satisfy the following conditions in order to avoid the complete invasion of the cavities by the liquid:

$$\left|\frac{dz}{dx}\right| = \left|tg(\pi - \theta_E)\right| = \left|tg(\theta_E)\right| \leq \left|\frac{dz}{dx}\right|_{\max}$$

Hence the condition $a\frac{2\pi}{\lambda} \geq \left|tg(\theta_E)\right|$

As the roughness r is given for low amplitudes and reasonable wavelengths by:

$$r = [\frac{dx^2 + dz^2}{dx^2}]^{1/2} \approx 1 + \frac{1}{2}\left|\frac{dz}{dx}\right|^2 \approx 1 + \frac{1}{4}\left|\frac{dz}{dx}\right|_{\max}^2$$

the boundary roughness r^* for the existence of air pockets is thus written as:

$$r^* = 1 + \frac{tg^2\theta_E}{4}.$$

Hence, for θ_E =120°, air pockets may appear for a boundary roughness of r^* =1.75.

Therefore, the simple interpretation is:

For a roughness $r<r^*$, the liquid follows the topography of the solid surface and the apparent contact angle is given by the Wenzel's equation (equation [2.1]). However, for a roughness $r>r^*$, air pockets appear and the apparent contact angle is deduced using the Cassie–Baxter equation (equation [2.10]).

This change, which is shown in Figure 2.7, can help in interpreting the notable wettability experiment carried out by Johnson and Dettre [JOH 64] with water on a wax surface showing variable and controlled roughness.

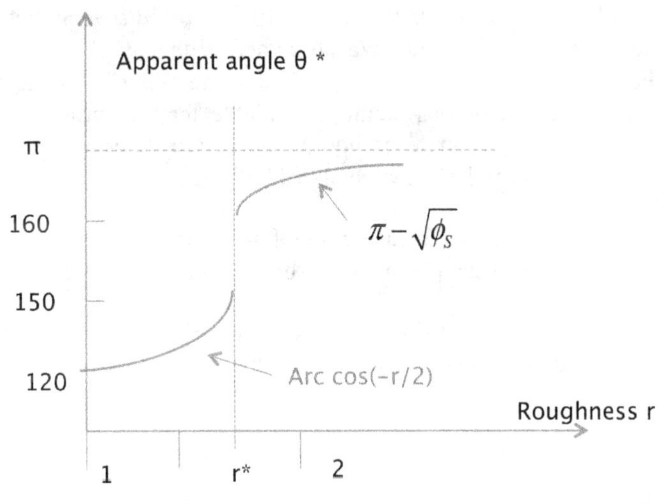

Figure 2.7. *Evolution of the apparent contact angle θ* for a surface model characterized by a Young's angle Θ E =120° and a boundary roughness of r*=1.75*

The development of micro or nanostructured surfaces has led to both experimental and theoretical studies on the extreme behavior of wettability (superhydrophilicity and superhydrophoby). The works cited in the reference list [WHY 08, BIC 02, PAR 12, SHI 96, BOR 07, DRE 11, HOU 08, HOU 11] discuss some of the results obtained.

2.5. Application

The Wilhelmy plate method shows the distribution of air pockets upon approach and the dispersion of air bubbles upon withdrawal.

The example presented below examines the wettability of a gas diffusion layer (GDL). This component is involved in the production of PEMFC fuel cells and is located between the active layer (mixture of carbon and a platinum-based catalyst) and the gas inlet device. The wettability of this component which carries and releases the water is an important parameter in the management of water in PEMFCs.

The study which was carried out used a GDL plate made of nafion-coated (PTFE) carbon fiber designed to improve the hydrophobicity of the device.

This study (Parry *et al.* [PAR 12, PAR 10]) involved several methods for measuring the wettability angle (Washburn and Wilhelmy plate). The results shown here are the measurements taken using the Wilhelmy plate method. This is one of the numerous methods available for measuring the surface tension and the contact angle of liquids which are described in reference works such as Adamson [ADA 90] and Wu [WU 82].

A sensor was used to measure the force during the immersion and withdrawal of the plate sample in the test liquid (in this case water).

Figure 2.8 shows the tensiometer used (Kruss K100 MK2) and the detail of the configuration of the liquid when the sample plate is withdrawn.

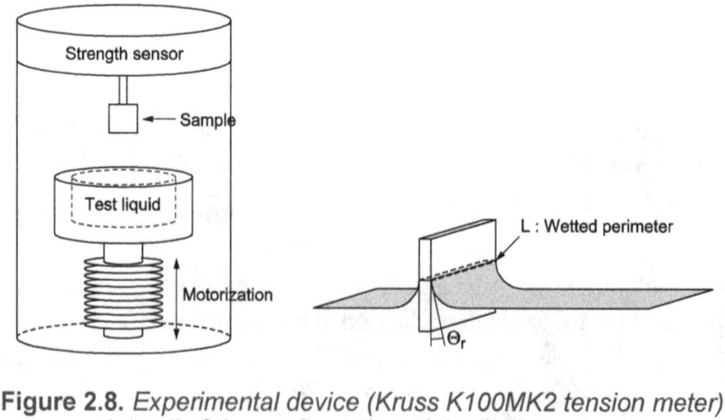

Figure 2.8. *Experimental device (Kruss K100MK2 tension meter) and detail of the configuration of the liquid during the withdrawal of the sample*

The sample is a plate with the dimensions of 25×25 mm^2 and a thickness of 230 µm. It is made of carbon fibers 10 µm in diameter. A preliminary treatment of the GDL by immersion is performed leading to a 10%wt charge with PTFE. Figure 2.9 shows a cross-section of the sample.

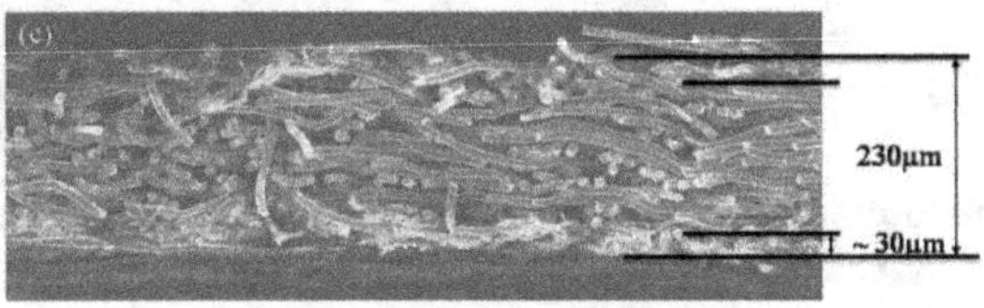

Figure 2.9. *SEM cross-section of the GDL with 10%wt of PTFE*

The force measured is given by the equation:

$$F = \gamma L \cos\theta + \pi$$

where L represents the wetted length and π Archimedes buoyancy.

The extrapolation to 0 of the immersion depth helps to eliminate the Archimedes buoyancy (dashed line). The contact angle can therefore be obtained, knowing the surface tension of the water.

An example of results is shown in Figure 2.10.

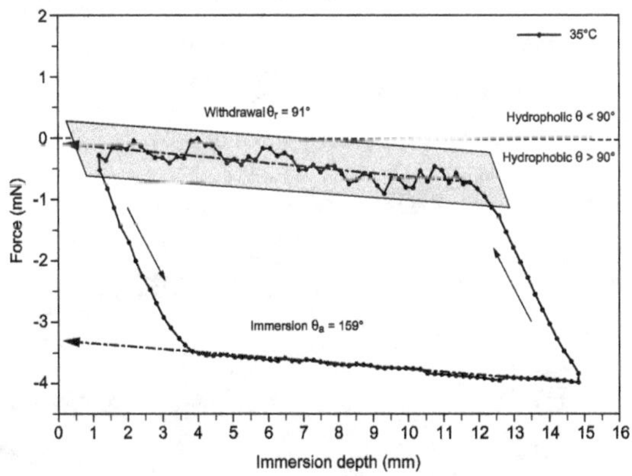

Figure 2.10. *Force measured during the immersion and withdrawal of the GDL plate in water at 35°C (continuous arrows)*

On this graph, it can be observed that upon immersion, the evolution of the force according to the immersion depth is perfectly constant since the roughness of the plate is smoothed by the air pockets. The approach angle is large, of the order of 160°.

Upon withdrawal, a higher value can be observed which corresponds to the release of the air pockets as soon as the contact line reaches the edge of a cavity on the plate. The contact angle obtained is lower, around 91°, and is close to that obtained by the Washburn impregnation method which characterizes the wetting angle inside the GDL.

3

Components of the Surface Energy

3.1. Overview

Fowkes hypothesis on writing surface tension as a sum of forces (dispersion and acid-base).

Van Oss and Good hypothesis where the surface tension is the sum of the Lifshitz–Van der Waals and the acid-base forces.

Surfaces can be classified into two categories:

– high energy surfaces where the bonding energy is of the order of the eV and which practically all liquids will wet. The bonds involved can be ionic, covalent and metallic. The corresponding surface energies are of the order of 500–3000 mJ/m^2;

– low energy surfaces where bonding energy is of the order of kT, or 2.52 10^{-2} eV at T = 20°C. This includes crystal molecules, polymers and organic liquids. The corresponding surface energies are of the order of 10–100 mJ/m^2.

In the following section, we will mainly focus on the second category, on which pioneering work was carried out by Fowkes [FOW 62] and Good [GOO 64].

The Fowkes hypothesis [FOW 62] is to assume that the surface tension of a liquid or a solid can be written as the sum of various contributions corresponding to the different intermolecular interactions:

$$\gamma = \gamma^d + \gamma^i + \gamma^p + \gamma^h + ... \qquad [3.1]$$

The indices correspond to different interactions:

d: London dispersion interaction;

p: Keesom dipole–dipole interactions;

i: Debye permanent-induced dipoles;

h: Interactions related to hydrogen bonds.

This concept is expanded upon by Fowkes [FOW 64] with the work of cohesion W_c and Dupré's work of adhesion W_a.

The work of cohesion is defined, in the same way as the work of adhesion, as the work corresponding to the separation of a pure liquid or solid into two distant parts.

$$W_c = W_c^d + W_c^i + W_c^p + W_c^h + \ldots \quad [3.2]$$

and

$$W_a = W_a^d + W_a^i + W_a^p + W_a^h + \ldots \quad [3.3]$$

Moreover, Fowkes proposes that for the interface 1-2 between two non-polar systems 1 and 2 where the London dispersion interactions are predominant:

$$W_a^d = \left[W_{c1}^d W_{c2}^d\right]^{1/2} = 2\left[\gamma_1^d \gamma_2^d\right]^{1/2} \quad [3.4]$$

Hence:

$$\gamma_{12} = \gamma_1 + \gamma_2 - 2\left[\gamma_1^d \gamma_2^d\right]^{1/2} \quad [3.5]$$

He also states that the interactions at an interface between two bulk phases only involve forces of the same nature, such as dispersion/dispersion, hydrogen bonds/hydrogen bonds, etc. Mixed forces, for instance, dispersion/polar, are not involved.

Later on, Fowkes [FOW 80] advocated retaining only the predominant terms in the expression of the work of adhesion and the free surface energy.

Moreover, he expands the notion of hydrogen bond to include it in the acid-base interactions, hence:

$$\gamma = \gamma^{d} + \gamma^{AB} \quad [3.6]$$

At about the same time, Good [GOO 64] used previous work carried out by Fowler and Guggenheim (1939), to suggest that the interaction energy at a flat interface can be separated into its different contributions, i.e. the dispersion, induction and dipole–dipole interactions.

These attractive interactions correspond to potentials in $1/r^6$ whose sum can be obtained by pair integration over all molecule positions.

The work of adhesion for a 1-2 system is written as follows:

$$W_a = -\Delta F_{12}^{a} = \frac{1}{16\pi l_0^{\,2}}\left[\frac{3\alpha_1\alpha_2 I_1 I_2}{2(I_1+I_2)} + \frac{\alpha_1\mu_2^2 + \alpha_2\mu_1^2}{2} + \frac{\mu_1^2\mu_2^2}{3kT}\right] \quad [3.7]$$

where α is the polarizability, I is the ionization energy, μ is the dipole moment, l is the separation distance between the two bodies varying from l_0 (equilibrium distance) to infinite, k is the Boltzmann constant and T is the temperature.

The final term of equation [3.7] corresponds to the interaction of free dipoles which is very low in the case of existing hydrogen bonds and is then replaced by a term specific to the hydrogen bond.

In the case of polar liquids or solids, Good [GOO 91] introduces an interaction parameter ϕ_{12} whereby:

$$\phi_{12} = \Delta F_{12}^{a} / \left[\Delta F_{11}^{c} \Delta F_{22}^{c}\right]^{1/2} \quad [3.8]$$

Given that the free energy of cohesion, for example of 1 pure, is written as:

$$\Delta F_{11}^{c} = 2\gamma_1$$

then $\gamma_{12} = \gamma_1 + \gamma_2 - 2\phi_{12}\left[\gamma_1\gamma_2\right]^{1/2}$ [3.9]

The experimental values of ϕ_{12} determined for various liquids vary between 0.5 and 1.15 with, for non-polar liquids, a value of ϕ_{12} =1 in accordance with Fowkes' proposal.

Further developments were introduced from 1985 onwards, based on the Lifshitz–Van der Waals theory (1956–1970). In this theory, the atomic structure is not taken into account and the forces of interactions between the macroscopic bodies are derived from macroscopic properties such as the dielectric constant and the refraction index.

This theory, set out clearly in Israelachvili's textbook [ISR 91], enables a calculation of the free surface energy taking account of all the electromagnetic interactions, whether they are due to temporary oscillating dipoles (γ_d), permanent dipoles (γ_p) or induced dipoles (γ_i).

Good [GOO 91] used this approach to calculate the interaction parameter ϕ_{12}.

The results of this calculation produce values of ϕ_{12} which are very close to 1 for various pairs of immiscible liquids where at least one of them does not display hydrogen bonds. The same result is obtained for the metallic mercury–gallium system.

Finally, for a solid with Lifshitz–Van der Waals (LW) interactions and acid-base interactions (AB), Van Oss, Chaudhury and Good [VAN 88] put forward the following expression which generalizes equation [2.6] proposed by Fowkes:

$$\gamma_i = \gamma_i^{LW} + \gamma_i^{AB} \qquad [3.10]$$

and for the work of adhesion:

$$W_a = W_a^{LW} + W_a^{AB} \qquad [3.11]$$

3.2. Molecular interactions and components of the energy

Ab initio calculations of the variations in the energy of molecules during their interaction.

This section summarizes Lee's analysis [LEE 93]. The variations in the energy of molecules, after and before their interactions, are calculated by an *ab-initio* Hartree–Fock method [KIT 76].

The Hartree–Fock procedure is self-consistent. The Schrödinger equation is resolved for an electron, assuming a potential created by an average distribution of the other electrons. A new orbital is obtained for the electron and the procedure is repeated for all the electrons.

The same cycle is repeated for the new orbitals obtained until convergence.

The result can be given as follows:

$$\Delta E_{\text{int}} = \Delta E_{el.st} + \Delta E_{pol} + \Delta E_{ex} + \Delta E_{ct} + \Delta E_{mix} \qquad [3.12]$$

– The electrostatic contribution $\Delta E_{el.st}$ corresponds not only to the Coulomb interactions between ionic charges but also to those between two dipoles (Keesom) or between an ionic charge and a dipole or between two quadrupoles, etc.

– The attraction component of polarization or induction ΔE_{pol}, consists of interactions between an induced dipole and an ionic charge or a permanent dipole (Debye).

– The repulsion component of exchange ΔE_{ex} arises from the exchange of electrons, over a short distance, between molecules A and B, caused by the overlap in the distributions of electrons between A and B. This reflects the Pauli exclusion principle and takes place at very short distances.

– The component of charge transfer ΔE_{ct} corresponds to the interactions between occupied and empty molecular orbitals between two molecules A and B. This contribution, which is also attractive, is the main contribution to the Lewis acid-base interaction (electron donor–acceptor interaction).

– The final contribution ΔE_{mix} corresponds to higher order interactions, including dispersion interactions. These correspond to fluctuation correlations in the distribution of electrons between neighboring molecules. They are present in every case, regardless of the other interactions.

The interactions known as Van der Waals or Lifshitz–Van der Waals (LW) interactions bring together the dispersion energy, Keesom dipole–dipole interactions included in the electrostatic term and the Debye dipole-induced dipole interactions in the polarization-induction term. Kitaura and Morokuma analysis uses different term groupings, which can lead to some confusion but can be overcome by the balancing of certain contributions.

Van der Waals interactions can be approximated using various theories as indicated above.

Historically, Van der Waals interactions were expressed using $-C/r^6$ interaction potential between molecules [DER 34, HAM 37]. Using techniques of summation, this approach enabled the evaluation of forces of interaction between surfaces with simple geometry (plane and sphere).

An example is provided at the end of this chapter.

3.3. The hydrogen bond

Similarities between a hydrogen bond and a Lewis acid-base reaction

The hydrogen bond is important in the case of aqueous solvents and solid-liquid interfaces. According to Pauling, this bond is partially covalent and partially ionic (or polar).

Using a self-consistent method, Morokuma and Kitaura [MOR 80] calculated the various energy terms involved in hydrogen bonding.

The various terms obtained are similar to those involved in the Lewis acid-base interactions.

The predominant components are the electrostatic and charge transfer terms. The electrostatic term is partially compensated by an exchange term, and the charge transfer term is mainly related to an electron exchange between the Lewis acid proton of the O-H group and the neighboring Lewis base with a lone pair such as: O|.

The theory of molecular orbitals can be applied, when the two molecules are close, to the construction of molecular orbitals using orbitals in their initial state.

The orbitals obtained are either bonding or antibonding. An occupied bonding level leads to stabilization of the system.

3.4. Lewis acid-base interactions

Mulliken and Klopman–Hudson analyses producing an interaction energy which is the sum of a Coulomb term and a term known as frontier orbitals.

Lewis acid-base interactions

$$A_{acid} + :B_{base} \rightarrow A:B_{complex} \quad [3.13]$$

combine not only hydrogen bonds and donor–acceptor pair interactions but also electrophile–nucleophile interactions. As previously discussed, these interactions involve ionic and covalent aspects.

Two theoretical analyses are available in the literature:

– The Mulliken approach [MUL 51] describes the state of the complex formed in terms of resonance between a non-bonded form (:B, A) and a form resulting from the transfer of an electron (B^{+}-A^{-}).

– The perturbation theory by Klopman and Hudson [KLO 67] uses methods from quantum mechanics where the wave function characterizing the perturbed system A:B is given by a linear combination of the wave functions which characterize the non-perturbed elements A and B.

The theory of perturbation applied to molecular orbitals (MO) [HUD 73] enables the interaction energy ΔE_{int} to be calculated.

The result can be given in the following form arising from the sum of a Coulomb term and a term known as a frontier orbital, ignoring the higher order terms.

$$\Delta E_{\text{int}} = -\frac{Q_N Q_E}{\varepsilon R} + \frac{2(C_N C_E \beta)^2}{E_{HOMO} - E_{LUMO}} \quad [3.14]$$

Q_N and Q_E are the charges of the nucleophile and electrophile sites, C_N and C_E are the coefficients of atomic orbitals N and E, β is the resonance or transfer integral $\beta = \langle 1|H|2 \rangle$ (Dirac notation), ε is the dielectric constant and R the distance between N and E, E_{HOMO} is the energy of the frontier orbital (the highest occupied level) of the nucleophile base and E_{LUMO} is the energy of the frontier orbital (the lowest empty level) of the electrophile acid. The coupling of the two levels, when they are close, creates a bond which allows the transfer of electrons, which is energetically favorable.

The origin of the second term in equation [3.14] can be easily understood in the theory of perturbations, considering the coupling between an empty level E_i^0 of a molecule i and the occupied level E_j^0 of a molecule j. The energetic stabilization obtained ΔE arises from the transfer of the electronic pair to the bonding level created by the coupling.

With an energetic stabilization as indicated in Figure 3.1:

$$\Delta E \simeq \langle i|H|j\rangle / \left[E_i^0 - E_j^0 \right]$$

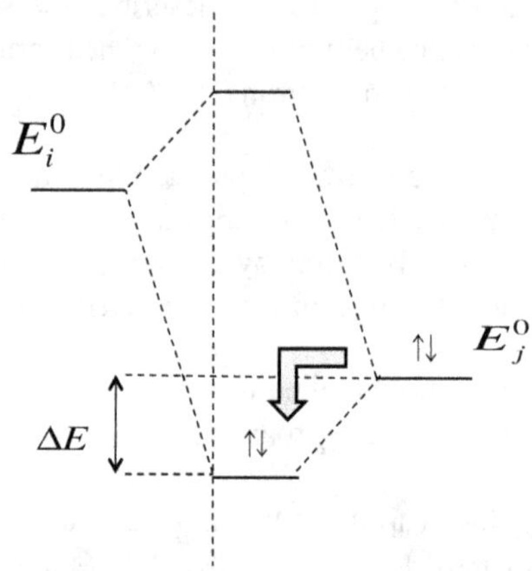

Figure 3.1. *Diagram of coupling between two electronic levels of molecules i and j. The creations of bonding and antibonding levels is only energetically favorable when the initial coupling takes place between an empty and an occupied level*

The Coulomb term of equation [3.14] is a long distance term inversely proportional to R.

However, the term of the frontier orbitals is short distance and inversely proportional to the energy difference between the HOMO and the LUMO belonging to the two molecules A and B, respectively.

This energy gap is identified as the absolute hardness of the HSAB principle [LEE 89].

When this gap is large, the Coulomb term in equation [3.14] is predominant. The interactions are controlled by the charges. This situation arises for highly polar compounds which are small in size and react over short distances. The donor element is highly electronegative (high ionization potential) and the acceptor has a low electron affinity.

However, when the energy gap is small, the term of the frontier orbitals (covalent) is predominant in equation [3.14]. This situation is promoted by the presence of components with low polar properties and reactive atoms with extended orbitals. The donor element has low electronegative properties and the acceptor has a high electron affinity.

The energetic stability obtained by the transfer of charge induced by the frontier orbitals is particularly significant in the case of strong acid-base complexes.

3.5. The effective components of the interaction energy

Simplification by term equalization, validation of Fowkes hypothesis.

The previous analysis shows that the interaction energy can be reduced to a four-term equation

$$\Delta E_{\text{int}} = \Delta E_{LW} + \Delta E_{ex} + \Delta E_{el.st} + \Delta E_{ct} \quad [3.15]$$

The Morokuma and Kitaura analysis shows that the terms ΔE_{ex} and $\Delta E_{el.st}$ can be approximately equalized and their sum is cancelled out.

Hence:

$$\Delta E_{\text{int}} \sim \Delta E_{LW} + \Delta E_{ct} \approx \Delta E_{LW} + \Delta E_{AB} \quad [3.16]$$

This validates Fowkes' proposal and the equation for the surface tension as a two-term summation:

$$\gamma \approx \gamma^{LW} + \gamma^{AB} \quad [3.17]$$

COMMENT 3.1.–

Van der Waals interaction forces between simple geometric surfaces (sphere and plane) are very clearly described by Israelachvili [ISR 91].

In a simple hypothesis (non-delayed and additive interactions), they involve a pair interaction potential $U(r)=-C/r^6$ and the so-called Hamaker constant A.

$$A = \pi^2 C \rho_1 \rho_2 \quad [3.18]$$

where ρ_1 and ρ_2 represent the densities of bodies 1 and 2 interacting across the vacuum (or the air).

This constant A has the order of magnitude $A = 10^{-19}$ J for interactions between condensed phases (solid or liquid).

Therefore, in the case of two flat surfaces 1 and 2 interacting across body 3 as indicated in Figure 3.2, the interaction energy W and the corresponding force are written as follows:

$$W(D) = -A_{132} / 12\pi D^2 \text{ and } F(D) = -A_{132} / 6\pi D^3 \quad [3.19]$$

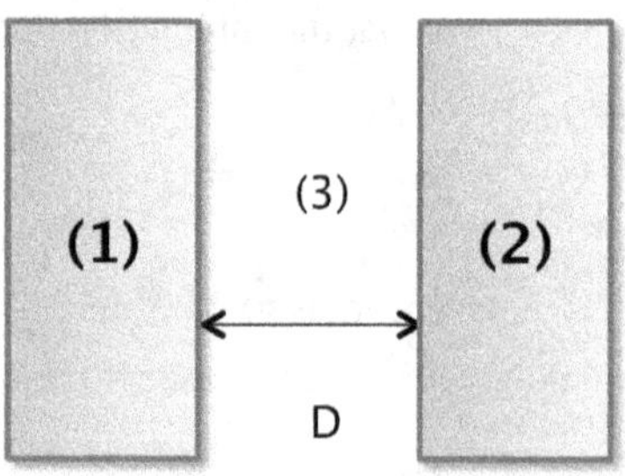

Figure 3.2. *Van der Waals attraction between two flat bodies 1 and 2 interacting across body 3*

The interaction energy is given by equation $W(D) = -A_{132} / 12\pi D^2$.

3.6. Application

Molecular bonding is implemented to obtain the silicon-on-insulator (SOI) device. Bonding is achieved by the Van der Waals forces and hydrogen bonds between the two surfaces of oxidized silicon. The measurement of bonding energy highlights the importance of hydrogen bonds at ambient temperature which are replaced by siloxane bonds at high temperature.

Molecular bonding, a commonly used process in the field of microelectronics, provides a good illustration of the role of surface forces which have been described in the previous sections. The creation of silicon-on-insulator (SOI) "wafers" aims to obtain a thin layer of high quality silicon (with few defects) on a layer of silicon oxide with a good interface, all of which is placed on a medium quality (and inexpensive) silicon substrate.

Different techniques have been used to reach this objective (Lasky [LAS 86], Shimbo [SHI 86], Maszara [MAS 88]).

Most recently, the "smart cut" process was developed by Bruel and the company SOITEC in collaboration with the laboratory LETI-CEA [BRU 95]. This process consists of extracting a thin layer of high quality silicon (A), due to a fracture caused by a controlled implementation of hydrogen. This layer is then bonded to a substrate of lower quality B as shown in Figure 3.3.

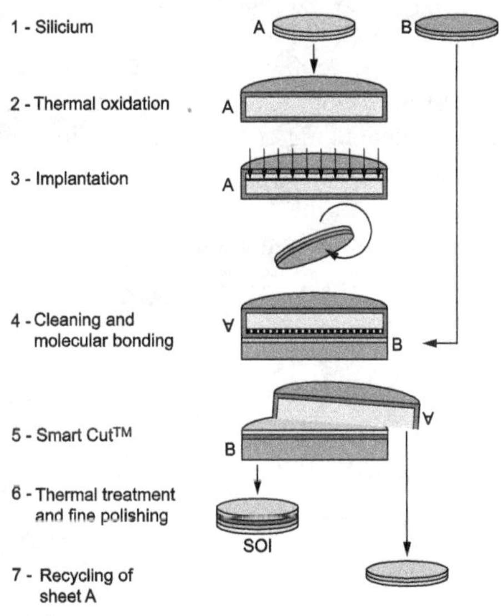

Figure 3.3. *Diagram of the "smart cut" process*

The key step in the molecular bonding between two oxidized surfaces is ensured by the action of the Van der Waals forces and the development of hydrogen bonds between the surface silanol compounds (SiOH). This bonding results from a genuine "bonding wave" which is even more effective when prior flattening of the surfaces has been carried out. This step is reversible.

At higher temperatures, the hydrogen bonds (Figure 3.4) are progressively replaced by covalent siloxane bonds (Si-O-Si) (non-reversible step).

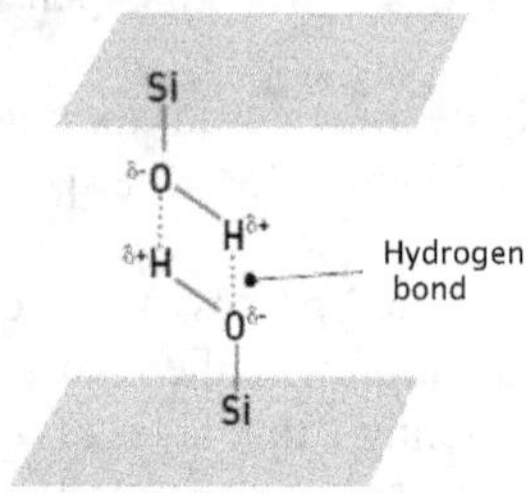

Figure 3.4. *Molecular bonding of two oxidized surfaces by hydrogen bonds*

The evolution in the nature of the bonds at the interfacial level is followed by the measurement of the bonding energy as a function of temperature. This energy is deduced from the Maszara plate experiment (1988) which is known in the adhesion community as the Boeing Wedge Test and was developed after WWII to study aluminum/polymer adhesion. (ASTM 3762-79/83).

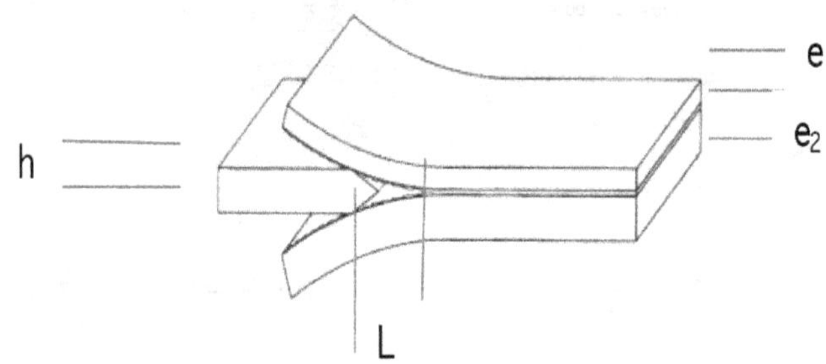

Figure 3.5. *Diagram of the wedge test*

A plate of thickness h is inserted between two silicon substrates, causing the development of a crack of length L (Figure 3.5).

In the case of two substrates with the same elastic modulus E and the same thickness e, a simple mechanical analysis can be used to calculate the adhesion energy G (linear elasticity calculation; Gillis and Gilman [GIL 64]).

The adhesion energy is reduced to two times the surface energy if the work of plastic deformation is not taken into account. In this calculation, only the deformation of the silicon is taken into account, ignoring the thin oxide–hydroxide layer on the surface.

$$G = 2\gamma_S = 3h^2 E e^3 / 16L^4 \quad [3.20]$$

Using the measurement of the length of the crack L obtained through various optical or acoustic techniques, the surface energy can be determined.

Figure 3.6 shows the measurements of the surface energy of a silicon wafer, after hydrophobic or hydrophilic treatments, as a function of the annealing temperature of the samples (Gosele [GOS 99], Ploβl [PLO 99]). At low temperatures, the samples of hydrophilic silicon show adhesion resulting from Van der Waals interactions and hydrogen bonds between the -Si-OH terminations.

Above 110–120°C, the elimination of water starts according to the following reaction:

-Si-OH + HO-Si-→ -Si-O-Si- + H_2O

The water absorbed and formed is likely to diffuse outside the interface across the native oxide layer to reach the bulk silicon and form hydrogen:

Si + $2H_2O$→ SiO_2 +$2H_2$

The hydrogen formed can either be eliminated along the defects at the interface, or will form microbubbles.

The surface energy changes only slightly between 200°C and 700°C. At higher temperatures, the surface energy increases significantly to values over 2 J/m^2. The native oxide becomes viscous from 800°C and can fill the microcavities between the surfaces which are not perfectly smooth.

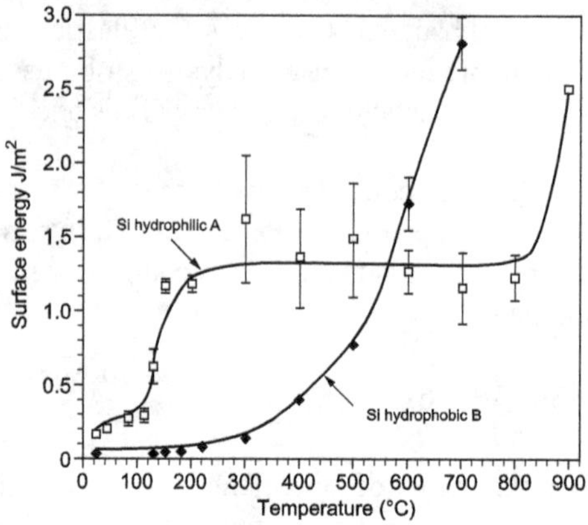

Figure 3.6. *Measurement of the surface energy determined by the plate method as a function of the annealing temperature of the sample (Gosele et al. [GOS 99] copyright 1999 with permission of Elsevier)*

The mechanisms discussed (formation of siloxane bonds (Stengl [3-24]), Gosele [GOS 99]) were extended by including mechanical aspects related to the deformation of roughness (Rieutord *et al.* [RIE 01]).

The hydrophobic silicon wafers are usually treated with diluted HF acid in order to eliminate the native oxide layer.

During heating, the desorption of hydrogen takes place from 400°C via the following reaction:

$$\text{-Si-H + H-Si-} \rightarrow \text{-Si-Si-} + H_2$$

The hydrogen molecules diffuse along the interface and along defects.

The surface energy then increases very rapidly from 600°C.

This type of study was conducted in detail by the groups of Moriceau and Rieutord [MOR 07]. Similarly, these groups carried out the detailed study of the interface closing mechanisms with temperature (lateral growth using roughness peaks Rauer [RAU 14]).

To conclude this example, it can be observed that at low temperatures, the difference between the surface energies of the two sample types shows the importance of the hydrogen bonds on the hydrophilic silicon.

4

The Acid-Base Component in the Work of Adhesion

4.1. Overview

Evaluation of the acid-base component in the work of adhesion using experimental measurements of the work of adhesion.

The acid-base component of the work of adhesion is generally evaluated according to the difference:

$$W_{12}^{AB} = W_{12}^{\exp} - W_{12}^{LW} \qquad [4.1]$$

The work of adhesion $W_{12}^{\exp}$ is deduced from wettability experiments which are detailed in Chapter 5.

The LW component, which is assimilated with the dispersive component W_{12}^{d}, is obtained by applying the geometric mean rule to the generally known values γ_1^d and γ_2^d:

$$W_{12}^{d} = 2\left[\gamma_1^d \gamma_2^d\right]^{1/2}$$

Values γ_1^d and γ_2^d are easily obtained through wettability experiments, e.g. between body 1 and an alkane. Indeed, taking into account the fact that the surface of alkanes only display dispersion interactions, the Fowkes rule stating that "interfacial interactions only involve forces of the same nature" enables the following relationship:

$$W_{1-alc} = \gamma_{alc}(1+\cos\theta_{1-alc}) = 2\left[\gamma_1^d \gamma_{alc}^d\right]^{1/2}$$

with $\gamma_{alc}^d = \gamma_{alc}$.

The surface energies of liquid alkanes are well-known. They are obtained by carrying out simple experiments to measure surface tensions such as the maximum bubble pressure method, the pendant drop method, or the Wilhelmy plate method [ADA 90, WU 82].

Table 4.1, taken from a review paper by Lavielle [LAV 89], shows surface tensions values for some alkanes and water.

Alkanes	γ alk (mJ/m^2	γ alk/water (mJ/m^2)
Pentane	16.2	51.4
Hexane	18.4	51.1
Heptane	20.3	51
Octane	21.3	51.0
Decane	23.4	51.0
Dodecane	25.3	51.2
Hexadecane	27.1	51.3
Water	γ water=72.6 ±0.2 mJ/m^2	γ^d water =21.6 mJ/m^2

Table 4.1. *Surface tensions of some alkanes and water [LAV 89]*

4.2. Use of the acid-base component

Fowkes' relationship between the acid-base component of the adhesion work and the enthalpy of reaction. Evaluation of the enthalpy of reaction using the Drago and Wayland empirical relationship.

Fowkes was the first to connect W_{12}^{AB} to the acid-base enthalpy of reaction per mole ΔH_{12}^{AB}, through the following relationship:

$$W_{12}^{AB} = -fN_{12}^{AB}\Delta H_{12}^{AB} \quad [4.2]$$

In this relationship, f is an enthalpy/free energy conversion factor which is close to 1. N_{12}^{AB} represents the number of active acid-base pairs per unit area across the interface 1-2.

The enthalpy of reaction can be calculated empirically using the Drago and Wayland equations [4.2] and [4.3]:

$$-\Delta H_{12}^{AB} = E_1^A E_2^B + C_1^A C_2^B \quad [4.3]$$

The parameters E and C measure the aptitude of a compound to participate in an electrostatic or a covalent bond respectively.

Using an arbitrarily chosen origin (iodine with $E^A = 1$ and $C^A = 1$), it is possible to build a scale of parameters E and C for different elements using experimental measurements of the enthalpy of reaction. These can be obtained using calorimetry or spectral measurements.

Fowkes and Mostafa [FOW 78] used this type of equation to correlate the adsorption of a PMMA base polymer dissolved in different solvents on an acid silica surface.

The acid-base properties of the various solvents used are characterized by the enthalpy of reaction with regard to the two reference reactants.

The adsorption of the PMMA on the silica surface is highest for the most neutral solvents and then decreases for the most acidic or base solvents.

In the first case, there is competition between the acid solvent and the acid surface for the adsorption of the PMMA. In the second case, the base solvent competes with the base polymer for adsorption over acid sites on the silica surface.

This type of study was extended to the adsorption of a PVC acid polymer on a base surface of calcium carbonate [KIN 87].

4.3. The Owens–Wendt approximation

The extended geometric mean rule.

Definition of the polar component of a surface.

Owens and Wendt [OWE 69], and later Kaelble and Uy [KAE 70], in line with Fowkes' initial work, suggest the following expression:

$$W_{12}^{n.d.} \approx W_{12}^{AB} \approx W_{12}^{p} = 2\left[\gamma_1^p \gamma_2^p\right]^{1/2} \qquad [4.4]$$

Hence an expression of the 1-2 interfacial tension:

$$\gamma_{12} = \gamma_1 + \gamma_2 - 2\left[\gamma_1^d \gamma_2^d\right]^{1/2} - 2\left[\gamma_1^p \gamma_2^p\right]^{1/2} \qquad [4.5]$$

This *extended geometric mean* rule has no theoretical basis, unlike the Fowkes hypothesis related to the dispersive component of the work of adhesion.

Indeed, the latter is justified by the Lennard-Jones expression of the mixed potential of interaction 1-2 using parameters (depth of the potential well, equilibrium distance 1-2) deduced from geometric means of parameters characteristics of interactions 1-1 and 2-2 in pure bodies (hypothesis on the identity of ionization energies of 1 and 2).

A slightly different rule known as the *harmonic mean* can be obtained using the hypothesis on the identity of the polarizabilities of 1 and 2 [WU 82].

The Owens–Wendt rule known as the *extended geometric mean* is frequently used in the literature thanks to its simplicity. It enables the user to easily obtain the polar component of a solid surface and thus to establish the total surface energy:

$$\gamma_S = \gamma_S^d + \gamma_S^p \qquad [4.6]$$

4.4. Van Oss–Good description

Definition of Lewis acid and base parameters which characterize a surface. Writing of the Lifshitz–Van der Waals and acid-base components of the interfacial tension between the substances 1 and 2. Possibility of negative values of the acid-base component which may lead to the solubilization of 1 into 2.

A combination rule was proposed by Van Oss *et al.* [VAN 86] to evaluate the acid-base component of the free surface energy.

The description set out uses the terminology of Lewis acid-base bonds, including the hydrogen bond as a special case.

A substance such as water has Lewis acid-base characteristics and can be considered bipolar. However, a substance which has a predominant characteristic is known as monopolar, e.g. $CHCl_3$ is a monopolar acid and ether oxide $(CH_3)_2 = O$ is a monopolar base.

The free surface energy includes a Lewis acid parameter γ^+ and a Lewis base parameter γ^- so that the free surface energy of a compound 1 is written as follows:

$$\gamma_1 = \gamma_1^{LW} + \gamma_1^{AB} = \gamma_1^{LW} + 2\left[\gamma_1^+ \gamma_1^-\right]^{1/2} \qquad [4.7]$$

Hence more precisely:

$\gamma_1^{AB} = 2\left[\gamma_1^+ \gamma_1^-\right]^{1/2}$ for a bipolar substance 1

and $\gamma_1^{AB} = 0$ for a monopolar or non-polar substance 1, i.e. without acid or base characteristics.

For an interface 1-2, this gives:

$$\gamma_{12} = \gamma_{12}^{LW} + \gamma_{12}^{AB} \text{ with} \qquad [4.8]$$

$$\gamma_{12}^{LW} = \gamma_1^{LW} + \gamma_2^{LW} - 2\left[\gamma_1^{LW} \gamma_2^{LW}\right]^{1/2} = \left[\sqrt{\gamma_1^{LW}} - \sqrt{\gamma_2^{LW}}\right]^2$$

$$\gamma_{12}^{AB} = 2\left[\sqrt{\gamma_1^+} - \sqrt{\gamma_2^+}\right]\left[\sqrt{\gamma_1^-} - \sqrt{\gamma_2^-}\right]$$

The work of adhesion is always written for an interface 1-2:

$$W_{12} = W_{12}^{LW} + W_{12}^{AB} \text{ with} \qquad [4.9]$$

$$W_{12}^{LW} = 2\left[\gamma_1^{LW}\gamma_2^{LW}\right]^{1/2}$$

$$W_{12}^{AB} = 2(\sqrt{\gamma_1^{+}\gamma_2^{-}} + \sqrt{\gamma_1^{-}\gamma_2^{+}})$$

As previously, the specific free energy of adhesion is linked to the work of adhesion by:

$$W_{12} = -\Delta F_{12}^{a} = -\Delta F_{12}^{LW} - \Delta F_{12}^{AB}$$

For example, the work of cohesion of a pure body 1 is written as follows:

$$W_{11}^{c} = -\Delta F_{11}^{LW} - \Delta F_{11}^{AB}$$ with

$$\Delta F_{11}^{LW} = -2\sqrt{\gamma_1^{LW}\gamma_1^{LW}} = -2\gamma_1^{LW}$$ and

$\Delta F_{11}^{AB} = -4\sqrt{\gamma_1^{+}\gamma_1^{-}}$ for a bipolar substance 1 and =0 for a monopolar or non-polar substance 1.

The advantage of this procedure is that it enables the description of the free surface energy using a three-parameter model, unlike the Owens–Wendt description which only involves two. Therefore, a more precise description of the free surface energy can be obtained.

Moreover, this procedure enables a negative value in the acid-base component of the free energy of interface 1-2.

Equation [4.8] shows that this result is obtained when:

$\gamma_1^{+} < \gamma_2^{+}$ with $\gamma_1^{-} > \gamma_2^{-}$ and vice versa.

In some cases, this situation can lead to an overall negative interfacial tension

$$(\left|\gamma_{12}^{LW}\right| < \left|\gamma_{12}^{AB}\right|)$$

This situation is mechanically unstable in the case of a liquid–liquid interface. However, for a solid–liquid interface, the situation can be mechanically metastable and observed over a given period of time.

The variation in the free energy associated with the transformation indicated in Figure 4.1 is written as follows:

$$\Delta F_{121} = -2\gamma_{12}$$

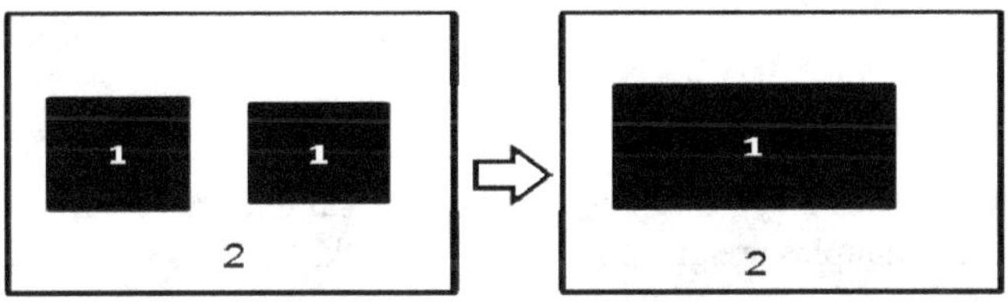

Figure 4.1. *Variation in the free energy associated with the condensation of two particles 1 immersed in a solvent 2*

A negative value of γ_{12} leads to a positive value of ΔF_{121}, i.e. there is stability in the dispersed state of the particles 1 in the liquid 2 (solubility).

Such a result is impossible in the Owens–Wendt analysis.

Indeed, in this case, the surface tension is written as $\gamma = \gamma^d + \gamma^p = \gamma^{LW} + \gamma^p$.

Therefore, for an interface 1-2:

$$\gamma_{12}^{LW} = [\sqrt{\gamma_1^{LW}} - \sqrt{\gamma_2^{LW}}]^2$$

$$\gamma_{12}^{p} = [\sqrt{\gamma_1^{p}} - \sqrt{\gamma_2^{p}}]^2$$

Hence $\gamma_{12} = \gamma_{12}^{LW} + \gamma_{12}^{p} \geq 0$.

The values of parameters γ^+ and γ^- required in the use of wettability measurements with test liquids are dependent on the choice of the reference liquid. The simplest and most natural choice is that of water in standard conditions, i.e.:

$$\gamma_w^+ = \gamma_w^-$$

Therefore, taking the values proposed by Van Oss in his latest book [VAN 06]

$$\gamma_W = 72.8\ mJ/m^2,\ \gamma_W^{LW} = 21.8\ mJ/m^2 \text{ leads to:}$$

$$\gamma_W^+ = \gamma_W^- = 25.5\ mJ/m^2 \text{ for water at 20°C.}$$

Table 4.2 shows some values of acid-base parameters of various liquids. Many other examples are given in Van Oss' book.

	γ	γ^{LW}	γ^{AB}	γ^+	γ^-
Water	72.8	21.8	51.0	25.5	25.5
Glycerol	64	34	30	3.92	57.4
Ethylene glycol	48.0	29	19.0	1.92	47.0
Formamide	58	39	19	2.28	39.6
Dimethyl sulfoxide	44	36	8	0.5	32
Chloroform	27.15	27.15	0	3.8	0
Diiodomethane	50.8	50.8	0		
Bromonaphtalene	44.4	43.5	0		

Table 4.2. *Free surface energy and acid base parameters (mJ/m^2) of some reference liquids at 20°C [VAN 06]*

5

Experimental Determination through Wettability Measurements

Wettability measurements are traditionally carried out on a flat surface using the sessile drop method. After a more or less thorough cleaning of the substrate depending on the subject being studied, a series (between 5 and 10) of identical liquid droplets (0.5 μl) are dropped onto the surface in order to determine the distribution of the contact angle. The objective is to obtain a nearly Gaussian distribution, whose central value determines the angle value.

The objective is to measure one or more equilibrium contact angles with different test liquids in order to calculate the surface energy of the solid being studied. Of course, the test liquids used should be as pure as possible to avoid problems associated with the adsorption of surface active impurities at various interfaces.

This indirect method is often the only possibility for obtaining the surface energy of a solid. The only direct method, known as the zero-creep method, can only be used for metallic systems at high temperatures with a high surface/volume ratio, in a gas-neutral atmosphere or in vacuum [BEN 83].

A distinction must be made depending on the type of surface being studied: surfaces with low surface energy (polymers) with $\gamma_S \leq 70-100\ mJ/m^2$, for which the spreading pressure π_e can be ignored and surfaces of high surface energy (metals and oxides) with $\gamma_S \geq 200-400\ mJ/m^2$, for which the spreading pressure must be taken into account.

5.1. One liquid method

This method is suitable for surfaces with low surface energy such as polymers. Application to a case of a non-polar surface, polyethylene, and a polar surface, polyaniline. Lavielle results.

The one liquid method is particularly suitable for the first situation (polymer materials). The measurements are carried out directly in the presence of liquid vapor, assuming that π_e is negligible, hence:

$$\pi_e = \gamma_S - \gamma_{SV} \approx 0$$

The work of adhesion is therefore written using Fowkes' notations:

$$W_{SL} = \gamma_{LV}(1+\cos\theta_E) = 2\sqrt{\gamma_S^d \gamma_{LV}^d} + W_{SL}^{AB} \quad [5.1]$$

a) In the case of a non-polar substrate such as polyethylene (PE), the acid-base component of the work of adhesion is zero because the interactions at an interface between two bulk phases only involve forces of the same nature.

Therefore, equation [5.1] is written as follows:

$$(1+\cos\theta_E) = 2\sqrt{\gamma_S^d}\,(\sqrt{\gamma_{LV}^d} / \gamma_{LV}) \quad [5.2]$$

The following graph (Figure 5.1) shows Lavielle results [LAV 89] obtained on the PE using various test liquids.

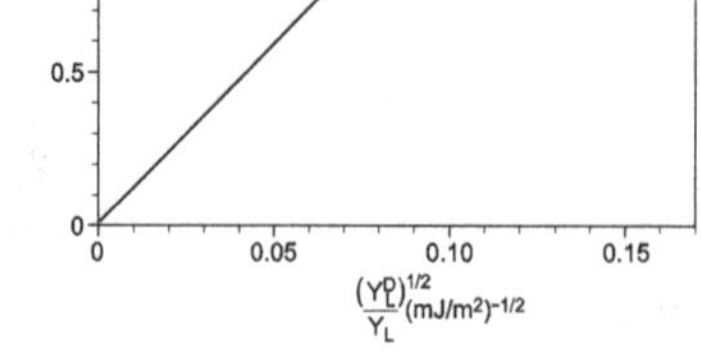

Figure 5.1. *Evolution of the contact angle obtained for different test liquids on a non-polar material, polyethylene. The straight line can be used to determine* γ_S^d *[LAV 89]*

The straight line gives: $\gamma_S = \gamma_S^d = 34.5\ mJ/m^2$.

b) In the case of a polar solid such as polyaniline, the experimental points obtained with various test liquids no longer form a straight line.

Indeed, there is an acid-base component which varies according to the test liquid used, as shown in Figure 5.2.

$$(1+\cos\theta_E) = 2\sqrt{\gamma_S^d}\,(\sqrt{\gamma_{LV}^d}/\gamma_{LV}) + W_{SL}^{AB}/\gamma_{LV} \qquad [5.3]$$

For a test liquid such as water, the acid-base component is shown by the difference:

$$\cos\theta_a - \cos\theta_b = W_{SL}^{AB}/\gamma_{LV}$$

The diagonal b) is obtained with an extra test liquid, mercury, which only displays dispersion interaction with the substrate.

The final result with water is thus: $\gamma_S^d = 41\ mJ/m^2$ and $W_{SL}^{AB} = 30\ mJ/m^2$.

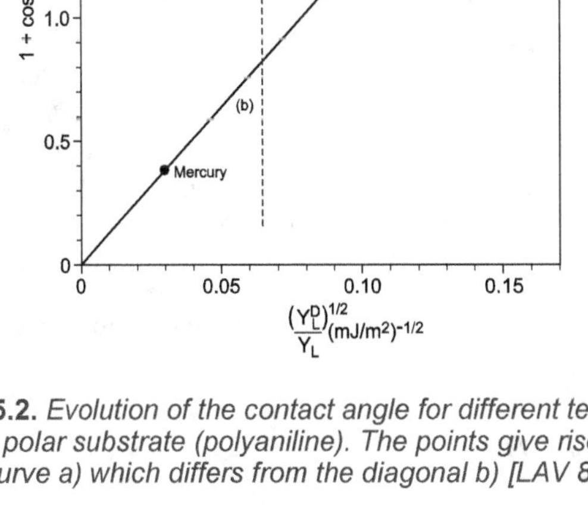

Figure 5.2. *Evolution of the contact angle for different test liquids on a polar substrate (polyaniline). The points give rise to a curve a) which differs from the diagonal b) [LAV 89]*

5.2. Two liquid method. Surfaces with high surface energy

Use of immiscible liquids (alkanes) to replace the vapor phase and avoid its adsorption on a solid surface. Establishing a master equation (Schultz).

In this situation, the spreading pressure π_e must be taken into account contrary to the case of surfaces with low surface energy. Indeed, in front of the droplet base, the vapor pressure of the liquid may lead to adsorption of water molecules causing a decrease in the surface energy of the solid.

To overcome this problem, Schultz, Tsutsumi and Donnet [SCH 77] suggest replacing the atmosphere surrounding the droplet with a liquid which is immiscible with the liquid droplet, usually an alkane (Figure 5.3).

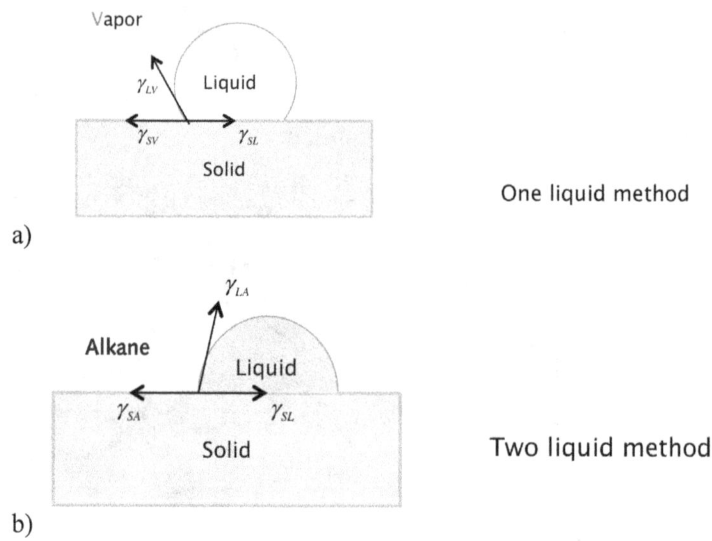

Figure 5.3. *a) Common one liquid configuration. Adsorption can take place from the vapor phase and modify the surface energy of the solid. b) Two liquid configuration. The alkane must be immiscible with the liquid being studied*

In the situation shown in Figure 5.3 (b), the condition of equilibrium is written as follows:

$$\gamma_{SA} = \gamma_{SL} + \gamma_{LA} \cos\theta_{SLA} \quad [5.4]$$

According to the Fowkes model and given that the alkane only displays dispersion interactions, thus $\gamma_A = \gamma_A^d$:

$$\gamma_{SA} = \gamma_S + \gamma_A - 2[\gamma_S^d \gamma_A^d]^{1/2} \quad [5.5]$$

Given that:

$$\gamma_{SL} = \gamma_S + \gamma_L - 2[\gamma_S^d \gamma_L^d]^{1/2} - W_{SL}^{AB} \quad [5.6]$$

The following general master equation can be obtained:

$$\gamma_L - \gamma_A + \gamma_{LA} \cos\theta_{SLA} = 2\sqrt{\gamma_S^d}[\sqrt{\gamma_L^d} - \sqrt{\gamma_A^d}] + W_{SL}^{AB} \quad [5.7]$$

The left-hand side of equation [5.7] shows experimentally accessible terms, while the right-hand side can be modified if the term $\sqrt{\gamma_L^d} - \sqrt{\gamma_A^d}$ is varied by changing the nature of the alkane used.

A linear representation of slope $2\sqrt{\gamma_S^d}$ is obtained which intercepts the origin at W_{SL}^{AB}.

5.3. Applications of the two liquid method

Two example applications are given, the first of which involves a thermal silica film on a wafer of monocrystalline silicon. The second example involves a sample of stainless steel AISI 304.

1) Thermal silica

Silica is studied using two test liquids: an aqueous phase of standard pH (5.8) and a basic aqueous phase (pH=11.4). The observed increase in the acid-base effect in the second case confirms the acidic character of silica. (Vittoz).

The samples are carefully cleaned with acetone before being dried with a neutral gas. The surface is analyzed with XPS (see Chapter 6). A layer of silica with a thickness of 6.5 nm is evidenced, which is covered with a carbon contamination layer 0.2 nm thick.

The stoichiometry of the layer of silica is obtained using the photopeak intensities corresponding to the oxidized silicon and oxygen, respectively: $ISi^{4+} / Io = 0.48$ [VIT 96].

The characteristic parameters of the alkanes used in the study are shown in Table 5.1. The data shown come either from the literature, or are obtained experimentally by the pendant drop method in air or in the corresponding alkane in the case of a water/alkane interface.

The measurements using the two liquid method were carried out with standard distilled water at pH=5.85 and in an aqueous base environment at pH=11.4.

It was experimentally verified (with the pendant drop method) that the pH of the aqueous solution had no influence on the liquid/alkane interfacial tension.

	Heptane	Octane	Decane	Dodecane	Hexadecane
Surface tension mJ/m^2					
Literature	20.3	21.3	23.4	25.35	27.1
Experimental	20.6 ± 0.3	21.9 ± 0.1			26.7± 0.1
Interfacial tension mJ/m^2					
Literature	51.0	51.0	51.0	51.2	51.3
Experimental	51.5 ± 1	51.5 ±1			51.2 ±1

Table 5.1. *Surface (alkanes) and interfacial (aqueous solution/alkane) tensions used in the two liquid method [VIT 96]*

Figure 5.4 shows the results which correspond to the application of equation [5.7].

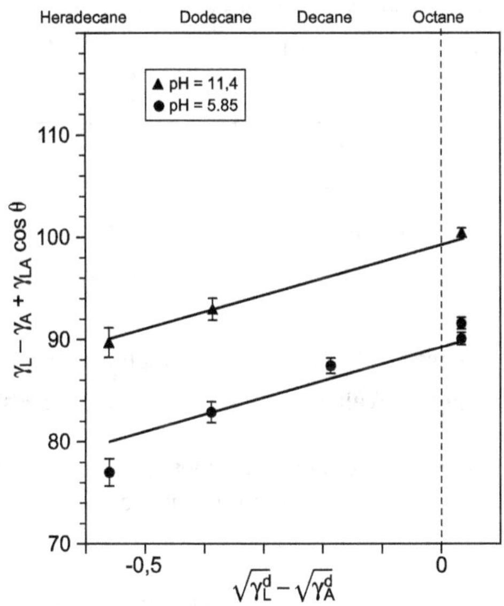

Figure 5.4. *Application of the two liquid method on a thermal SiO2 surface [VIT 96]*

The application of equation [5.7] to aqueous solutions with pH 5.85 and 11.4 produces two parallel straight lines with a slope $2\sqrt{\gamma_S^d}$ intercepting the origin at W_{SL}^{AB}.

The uncertainty on the γ_S^d value, which is determined using a slope, is significant (~15–20%). The uncertainty observed for W_{SL}^{AB} is much lower (~5%).

The conclusion of this study is that the pH of the aqueous solution does not significantly modify the dispersive component of the surface energy of the silica being studied. However, the acid-base component of the solid/liquid interface increases significantly with the basicity of the aqueous solution.

This result confirms the acid nature of the SiO_2 surface.

The numerical results are shown in Table 5.2.

Surface	γ_S^d mJ/m^2	W_{SL}^d mJ/m^2	W_{SL}^{AB} pH=5.85 mJ/m^2	W_{SL}^{AB} pH=11.4 mJ/m^2
SiO_2	85 ± 15	86.8	89	100

Table 5.2. *Energy parameters of an SiO_2 surface as measured by the two liquid method for aqueous solutions of pH 5.85 and 11.4 [VIT 96]*

The spreading pressure $\pi_e = \gamma_S - \gamma_{SV}$, which exists in the case of a one liquid experiment may be evaluated using an additional measurement.

The contact angle of a standard droplet of distilled water of pH 5.85 measured in a one liquid experiment on the surface of SiO_2 is equal to 41 ± 1 degree.

The relationship connecting the total solid–liquid work of adhesion to the liquid–vapor tension is written as follows (the Young Dupré equation):

$$W_{SL} = \gamma_{LV}(1 + \cos\theta) + \pi_e$$

The total solid–liquid work of adhesion is the sum of the dispersive and acid-base components obtained from the previous experiment and indicated in Table 5.2.

Knowing the value of γ_{LV}, the spreading pressure obtained is written as:

$$\pi_e = 48 \pm 10 mJ / m^2 .$$

This value is not negligible and justifies the use of the two liquid method for high energy surfaces.

2) Stainless steel AISI 304

The surface of austenitic stainless steel is studied following cleaning by Ar plasma at ambient temperature or after baking at 200°C. The test liquid is an aqueous solution with pH 10.7. The low acid-base effect obtained for the heated state can be linked to the enrichment of the surface by iron oxide (XPS) causing an increase in the base nature of the surface (Dubois).

The second example presented here involves an industrial austenitic alloy with a mean composition of Fe-Cr18%-Ni10% with various quantities of impurities such as Si and C. There are usually two types of surface state obtained following metallurgical operations: bright annealed obtained after annealing at high temperature (810°C) in a H_2/H_2O atmosphere or annealed-and-etched (annealed in a slightly oxidizing atmosphere followed by surface etching [BAR 90]).

A passive surface film (Fe and Cr oxy hydroxide) gives to this steel its ability to resist corrosion.

The study presented below involves bright annealed samples.

In their "as received" state, the samples studied show a significant carbon contamination caused by the production process and their exposure to the atmosphere. It is therefore recommended to decontaminate the surface being studied with a soft argon plasma treatment (10 min, 50 W). This treatment cleans the surface without modifying the surface Fe and Cr oxy hydroxide layer [DUB 00].

It was applied to all the samples in this study, whether they were studied at ambient temperature, or heated for 10 min at 200°C.

XPS analysis at normal and grazing incidence (see Chapter 6) can be used to evaluate the level of carbon contamination (Table 5.3).

Stainless steel AISI 304	Contaminated	Decontaminated	Decontaminated+ heated
Contamination thickness (nm)	2.1	0.4	0.4

Table 5.3. *Thickness of carbon contamination on the "as received" state, decontaminated with plasma Ar (50W, 10min) and decontaminated and heated at 200°C for 10 min [DUB 00]*

These samples were studied using the two liquid method with alkanes whose surface properties are shown in Table 5.1.

Several aqueous solutions with varying pH were used [DUB 00].

The results obtained for aqueous solution with pH 10.7 are shown in Figure 5.5.

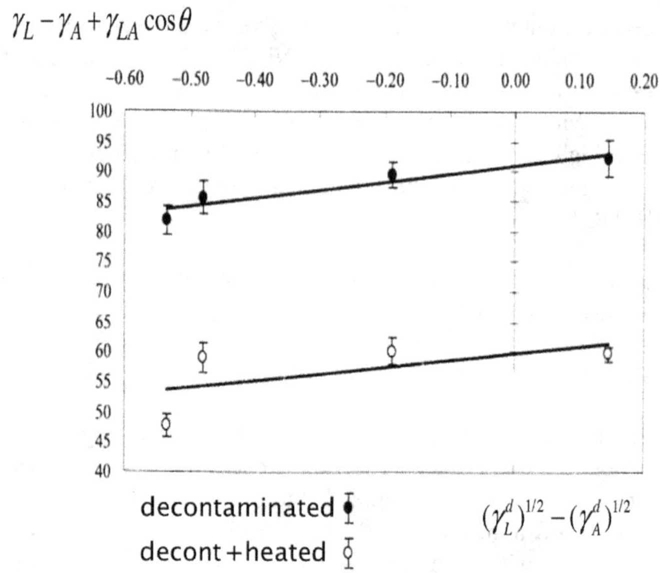

Figure 5.5. *Two liquid method applied to a bright annealed 304 stainless steel surface. The aqueous solution used is pH=10.7. Two surface states are studied: decontaminated (Plasma Ar 10 min-50 W) and decontaminated and heated (10 min-200°C).*

In both cases, a clearly linear representation is obtained as a function of $\sqrt{\gamma_L^d} - \sqrt{\gamma_A^d}$.

The common value of γ_S^d deduced from the slope is 40 ± 15 mJ/m^2.

The ordinates at the origin correspond to the acid-base component of the solid/liquid work of adhesion. The values obtained are respectively:

– decontaminated state: $W_{SL}^{AB} = 90 \pm 1$ mJ/m^2;

– heated and decontaminated state: $W_{SL}^{AB} = 59 \pm 1$ mJ/m^2.

For a basic aqueous solution, the work of adhesion W_{SL}^{AB} decreases for a heated surface, i.e. the base nature of the surface increases with moderate heating. The change in the surface's chemical composition caused by the heat treatment is shown by an increase in the number of base sites.

This observation should be correlated with the enrichment of the surface in Fe oxy hydroxide measured by XPS.

The change in the surface composition should also lead to a modification of γ_S^d.

This value is likely to be low and included in the significant uncertainty indicated above.

5.4. Comparison between Owens–Wendt and van Oss–Good methods

Both methods are applied to a cleaned ferritic steel surface, tested either at ambient temperature or after baking at 250°C or 800°C. Apart from the small differences observed in the total surface energy deduced from the two methods, only the van Oss method indicates an increase in the base nature of the surface with temperature, linked to an enrichment of the surface with oxide.

The last example presented here examines a comparison between Owens–Wendt and Van Oss–Good analyses carried out on ferritic steel as part of a study on bio-adhesion [GRO 13].

The steel studied is ferritic, with an industrial surface state known as annealed and etched (annealed in an oxidizing atmosphere, followed by surface etching).

Various grades of steel were studied, but only the results for K41, the composition of which is given in Table 5.4, are presented here.

Grade	C	Si	Mn	Cr	Mo	Ti+Nb
K41 %	0.015	0.60	0.30	17.8	0.65	0

Table 5.4. *Composition (wt %) of the ferritic grade studied*

In addition to the untreated (reference) surfaces, samples with various heat treatments were studied:

– 250°C for 5 min or 180 min;

– 850°C for 5 min or 180 min.

The surface being studied undergo a relatively thorough cleaning process:

– ultrasonic cleaning for 5 min in a 50% ethanol 50% acetone solution;

– paper wiping and air drying;

– plasma Argon 10 min 50 W;

– air aging 1 h before wettability measurements.

This air aging step causes a slight recontamination of the surfaces studied but allows a better control of the time between the end of the plasma treatment and the wettability measurements. This time must be kept constant to allow comparisons at (low) constant contamination.

The wettability experiments were carried our using the one liquid method with various test liquids.

This configuration was chosen to limit the number of experiments. The spreading pressure π_e is not taken into account, meaning that the work of adhesion W_{SL} evaluated by the Owens –Wendt and Van Oss methods will be underestimated. The components of the surface energy to be calculated are only indicative. This study focuses on the change in the components of the surface energy after carrying out heat treatments.

The characteristics of the four test liquids, which are shown in the table below, are taken from Van Oss's book [VAN 06].

Water				Formamide				Diiodomethane				Glycerol			
γ	γd	γ+	γ–	γ	γd	γ+	γ–	γ	γd	γ+	γ–	γ	γd	γ+	γ–
72.8	21.8	25.5	25.5	58	39	2.28	39.6	50.8	50.8	0	0	64	34	3.9	57.4

Table 5.5. *Characteristics of the four test liquids used in the study of two ferritic steel surfaces*

3) Analysis using the Owens–Wendt method

The solid–liquid work of adhesion is written as follows: $W_{SL} = \gamma_L(1+\cos\theta) = 2\left[\gamma_L^d \gamma_S^d\right]^{1/2} + 2\left[\gamma_L^p \gamma_S^p\right]$

$$\text{Hence: } \frac{\gamma_L(1+\cos\theta)}{2\sqrt{\gamma_L^d}} = \sqrt{\gamma_S^d} + \sqrt{\gamma_S^p}\left[\sqrt{\gamma_L^p} / \sqrt{\gamma_L^d}\right] \qquad [5.8]$$

The following figure shows this representation in the case of an annealed and etched-K41 surface following cleaning and 1 h aging.

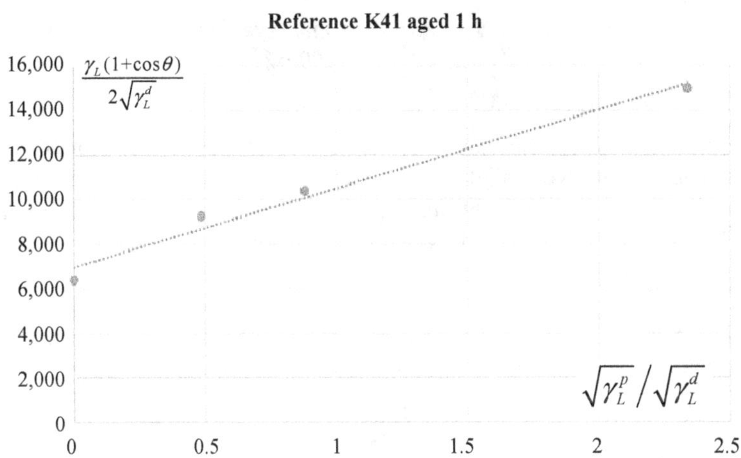

Figure 5.6. *Application of equation [5.8] to a surface of K41 ferritic steel aged for 1 h after plasma treatment*

This type of treatment is applied to various samples which have undergone heat treatments.

A selection of the results obtained is shown in Table 5.6.

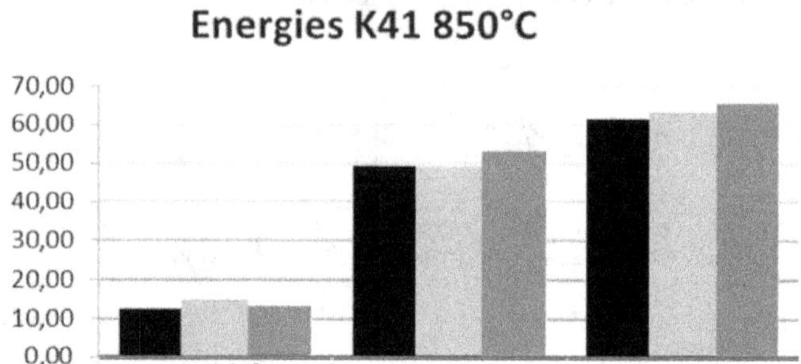

	γp mj/m²	γd mj/m²	γ mj/m²
K41reference	12,32	48,86	61,18
K41 5min	14,29	48,86	63,15
K41 180min	12,74	52,71	65,45

Table 5.6. *Owens–Wendt method: components of surface energy measured after heat treatment at 850°C for the K41 grade ferritic steel*

The results show the values of the surface energy components of grade K41. They show a very slight increase in the dispersive component with the duration of the heat treatment, linked to the enrichment of the surface mainly by chromium oxide observed by XPS. The polar component does not show any clear changes.

The results obtained for a treatment temperature of 250°C are very close to those shown in Table 5.6, and show no significant change with the duration of the treatment.

4) Analysis using the Van-Oss and Good method

Without taking the spreading pressure into account once again, the solid–liquid work of adhesion is written as follows:

$$W_{SL} = 2\left[\gamma_S^{LW}\gamma_L^{LW}\right]^{1/2} + 2\left[\gamma_S^{+}\gamma_L^{-}\right]^{1/2} + 2\left[\gamma_S^{-}\gamma_L^{+}\right]^{1/2} \qquad [5.9]$$

Equation [5.9] is used with the first three test liquids in Table 5.5. This slightly dubious procedure (three test liquids for three unknowns) is supplemented with a second group of three test liquids including glycerol instead of formamide. Reasonable convergence between the results is obtained.

The results, which are only indicative, are shown in Table 5.7.

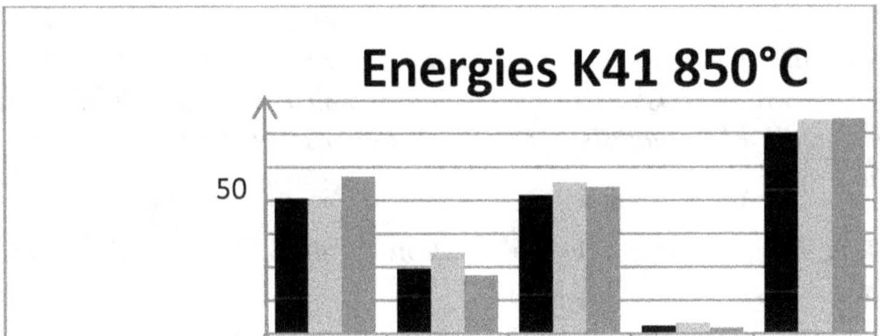

	γLW	γAB	γ-	γ+	γS
■ k41 0min	40.60	19.47	41.57	2.28	60.07
■ k41 5min	39.86	24.30	45.35	3.25	64.16
■ k41 180min	47.07	17.36	43.98	1.71	64.43

Table 5.7. *Van Oss and Good method; components of the surface energy measured after heat treatment at 850°C for the K41 grade ferritic steel*

The results obtained show that the duration of the heat treatment has no significant effect on the total value of γ_S which is comparable to that obtained in Table 5.6.

However, an increase in the base component of the surface γ_S^- can be observed, correlating with a very weak acid component γ_S^+. These results indicate a change in the surface toward a monopolar base nature.

This type of evolution was observed for grade K44 which has the same composition of Cr with minor additional elements such as Mo and Ti, Nb.

The comparison between the Owens–Wendt and Van Oss–Good methods produce overall values of γ_S which are slightly different, due to the spreading pressure hypothesis and the limited number of test liquids used.

Nevertheless, only the Van Oss–Good method shows a clear evolution in the acid-base nature of the surface after heat treatment, linked to the quantity of chromium oxide present at the outermost surface.

The final two examples on austenitic and ferritic steel show that heat treatment which is associated with an enrichment of oxides on the surface, indicated by XPS analysis, leads to an increase in the base nature of the surface.

This point is important for controlling the adhesion of heat-treated surfaces.

6

Acid-Base Properties of Surfaces: Experimental Approaches

6.1. Overview

Chapter 3 showed the importance of the contribution of acid-base interactions to the work of adhesion. Several expressions enabling the evaluation of the acid-base component were formulated. They are based on the measurements of the free surface energy, but in general only provide an overall mean measurement (on a macroscopic scale) of acido-basicity. For some applications, such as bonding or catalysis for example, it is sometimes more valuable to have a more localized idea (on an atomic or molecular scale) of the acid-base characteristics of the surface in order to determine the potentially reactive sites.

This chapter present a few experimental methods which can be used for evaluating the acid-base properties of solid surfaces, focusing on the characterization of flat surfaces, which is less documented than that of particles. Since a surface is generally made up of different chemical species, its acido-basicity depends on the density, strength and accessibility of the various sites. It can therefore vary according to the measurement method used. In practice, the choice of a method often depends on the preferred scale of characterization (general or local).

6.2. General methods

6.2.1. *Wettability*

The simplicity of implementing the wettability method and the large amount of literature dedicated to it since the work of Young in 1805

[YOU 05] mean that it is the most commonly used technique in a number of domains. Measurements are carried out on a flat substrate using the one or two liquid sessile drop method. Both techniques can be used to measure the work of adhesion and to deduce the value of the acid-base component by subtracting the dispersive component. The one liquid method is suited to materials with a low surface energy (polymers) whilst the two liquid method works best with solids with high surface energy (metals, oxides, etc.). These two methods, as well as the theories on which they are based, are described in Chapters 1–5, to which the reader may refer.

When molecular groups present at the surface of a solid can be ionized (e.g. hydrated groups Me-OH at the surface of a metal), the combination of wettability and electrocapillarity methods can be used to ascertain the acid-base nature of the surface. This approach is described in Chapter 7.

6.2.2. *Electrokinetic method*

Electrokinetic measurements give access to the properties of charged particles/solution interfaces. Initially developed for solid oxide particles, this approach was then expanded to flat surfaces.

6.2.2.1. *Double electrochemical layer*

A particle or a flat surface immersed in a fluid acquires an induced surface charge linked to the ionic adsorption or the ionization of the charged groups on the surface. As a consequence, counter ions are accumulated close to the surface, giving rise to an interface between the solid and the liquid formed of two layers of reverse polarity known as the "double electric layer". In 1853, Helmholtz suggested modeling this double layer with a condenser whose capacity depends solely on the dielectric constant of the electrolyte and the thickness of the double layer [HEL 53] (see Figure 6.1). The potential within this layer therefore varies linearly from the value Ψ_0 of the charged surface to the zero value corresponding to the electrolyte (Figure 6.1).

Gouy in 1909 [GOU 09] and Chapman in 1913 [CHA 13] improved this model by introducing a distribution of charges within the double layer (diffuse layer). Therefore, the potential within this layer varies exponentially from the value Ψ_0 to zero (Figure 6.2). The thickness of the diffuse layer is known as the Debye length.

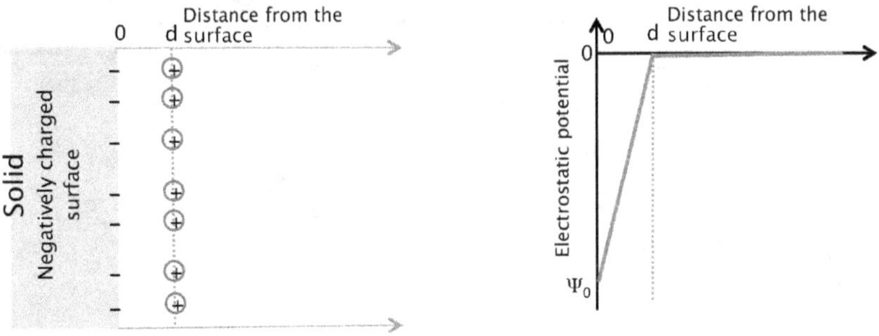

Figure 6.1. *Helmholtz model of the interface between a negatively charged surface of potential Ψ_0 and an electrolyte solution*

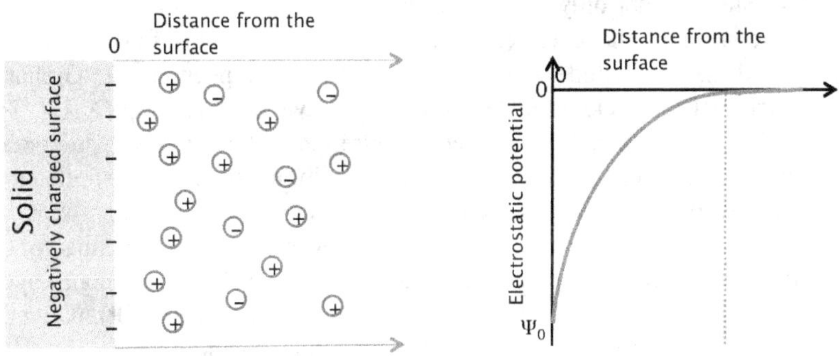

Figure 6.2. *Gouy–Chapman model of the interface between a negatively charged surface of potential Ψ_0 and an electrolyte solution*

The Gouy–Chapman model considers the ions as discrete points and the solvent as a dielectric continuum. In order to take into account the finite size of the ions, Stern suggested in 1924 breaking down the solid–electrolyte interface into two parts [STE 24]: a compact layer exempt of ions, known as the Stern layer, with thickness d corresponding to the minimum approach distance of the ions and modeled by a condenser of constant capacity C, then a diffuse layer with identical characteristics to the Gouy–Chapman layer (see Figure 6.3). The plane separating the compact layer from the diffuse layer is known as the Stern plane. The potential thus varies linearly from the value Ψ_0 to the value Ψ_d, then exponentially to zero.

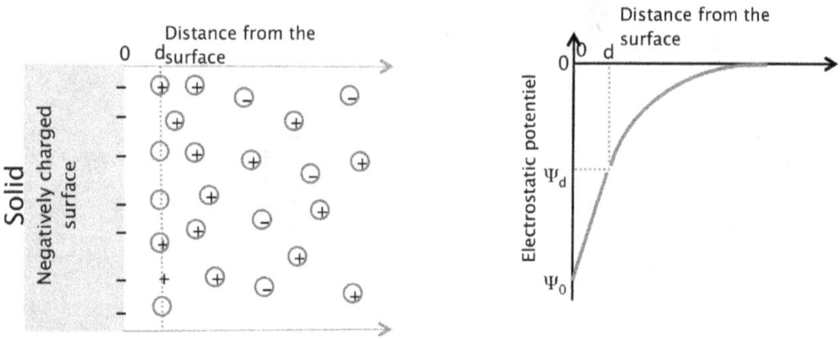

Figure 6.3. *Stern model of the interface between a negatively charged surface of potential Ψ_0 and an electrolyte solution*

The Stern model only considers the Coulomb interactions and does not take into account the chemical affinity which some ions may have with surface sites. To include these specific adsorption phenomena, Grahame suggested in 1947 dividing the Stern layer into two sublayers [GRA 47]. The model thus obtained has three layers (Figure 6.4). The first layer, in contact with the surface, contains specifically adsorbed ions. It can be assimilated to a condenser with constant capacity C_i. The second layer contains hydrated ions and can be assimilated to a second condenser with constant capacity C_e. The third layer corresponds to the diffuse layer. The first and second layers are separated by a plane known as the inner Helmholtz plane (IHP), while the second and third layers are separated by the outer Helmholtz plane (OHP), similar to the Stern plane shown in the previous model. Therefore, the potential varies linearly between Ψ_0 and Ψ_i, then between Ψ_i and Ψ_d, and finally exponentially to zero.

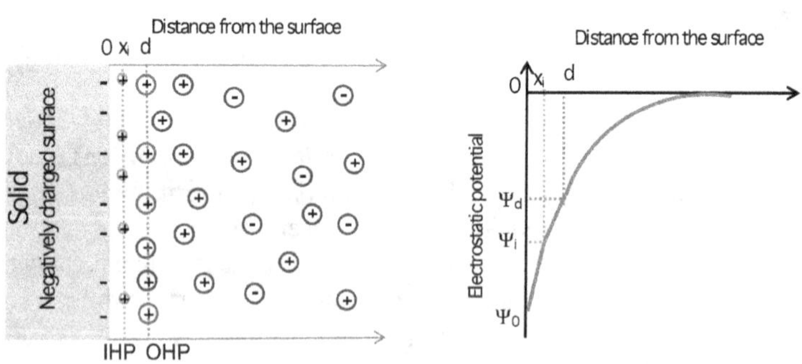

Figure 6.4. *Grahame model of the interface between a negatively charged surface of potential Ψ_0 and an electrolyte surface*

6.2.2.2. *Zeta potential – point of zero charge*

Applying an electric field in a suspension containing charged particles induces their movement. Within the double layer, the ions which have a specific affinity with its surface move with the particle, while the other ions will move away from the surface. These differences in behavior give rise to a hydrodynamic shear plane, which is generally very close to the outer Helmholtz plane. In this plane, the potential Ψ is equal to the electrokinetic potential or potential ς (zeta). Its value is closely dependent on the concentration of ions determining the solution potential. The concentration at which it cancels itself out defines the point of zero charge (PZC), i.e. the value at which the electrical surface charge is zero. In most cases, the ions which determine the potential are the ions H^+/OH^-. The value of the PZC thus corresponds to the pH value of the solution leading to a zero value of the potential ς. When the pH of the solution is lower than the PZC of the solid, the surface is positively charged. For pH values higher than the PZC, it is negatively charged. In other words, the surface becomes more acidic as its PZC gets lower, and more basic as it gets higher. The presence of specifically adsorbed ions can move the pH value for which the potential ς is cancelled out. The obtained value then defines the surface's isoelectric point (IEP).

6.2.2.3. *Measurement of the streaming potential*

Electrokinetic methods can be used in order to determine the potential ς and, in particular, the PZC or the IEP. Initially developed for oxide powders, these methods are commonly used to study colloids and biological cells. The most widely used method is electrophoresis, which is the measurement of the speed of displacement of the particles under the effect of an electric field. Any reader interested in the theory and implementation of these methods to determine the electrical properties of particle–liquid interfaces can refer to a recent review paper published under the auspices of the IUPAC (International Union of Pure and Applied Chemistry) [DEL 07]. In this book, we are only focusing on the extension of these measurement to flat surfaces, as described by van Wagenen and Andrade in 1980 [VAN 80].

Determining the PZC of flat surfaces relies on the measurement of the streaming potential. It principle is shown in the diagram in Figure 6.5.

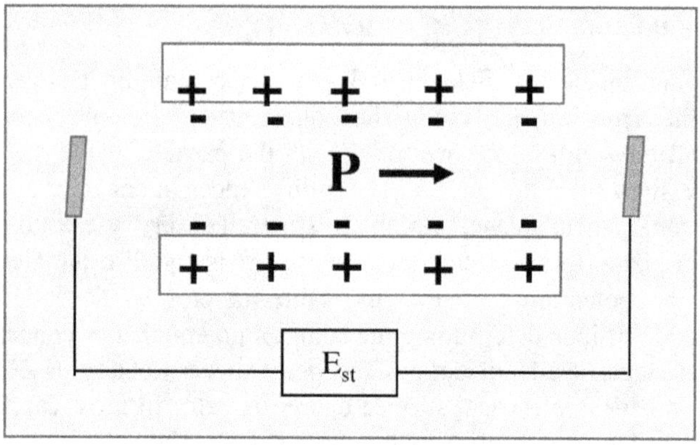

Figure 6.5. *Principle of the measurement of the streaming potential E*

Two plates made of the solid being studied are placed parallel to each other in an electrolytic cell, with the surfaces to be characterized facing each other. The electrolyte solution is moved in parallel to the plates under the effect of a pressure difference P of the order of 10–500 mbar through the cell. The streaming between the plates moves the mobile charges in the solution, in particular those in the Gouy layer known as counter-charges, while the charges at the center of the solution do not generate any current (electroneutrality). The excess of these counter-charges creates a streaming or convection current and a difference of potential E, known as the streaming potential, which appears at the edges of the plates. Since the liquid is a conductor, a conduction current opposite to the previous current, also appears. At equilibrium, these two currents are equal. The appearance of a relative speed due to the movement of counter-charges creates a shear plane or hydrodynamic cut in the double electric layer in which the streaming is stationary and whose potential is the potential ζ. The streaming potential E and the potential ζ are linked by the following expression:

$$\zeta = \frac{\eta}{\varepsilon\varepsilon_0}(\lambda_{sol} + \frac{L_{surf}}{b})\frac{E}{P} \quad [6.1]$$

where η, ε, ε_0 and λ_{sol} represent the viscosity, relative and absolute permittivities and the conductivity of the solution respectively. L_{surf} represents the specific conductance of the surface, b the half-distance between the plates, E the potential difference at the edges of the plates and P the pressure applied.

By making a suitable choice of the characteristics of the solution, it is possible to establish conditions so that L_{surf}/b is negligible compared with λ_{sol}. Equation [6.1] can therefore be simplified and becomes:

$$\zeta = \frac{\eta}{\varepsilon\varepsilon_0}\lambda_{sol}\frac{E}{P} \qquad [6.2]$$

In practice, the streaming potential often depends on the direction of the flow. It is therefore preferable to calculate the potential ζ by varying the pressure P in both flow directions and to use the slope $\Delta E/\Delta P$ obtained by linear regression. The potential ζ is therefore given by the Helmholtz–Smoluchowski equation:

$$\zeta = \frac{\eta}{\varepsilon\varepsilon_0}\lambda_{sol}\frac{\Delta E}{\Delta P} \qquad [6.3]$$

It should be noted that this relationship is only valid in the case of a laminar and stationary flow. By measuring ζ for various values of the solution pH, it is then possible to draw the curve ζ=f(pH). The intersection of this curve with the origin of the axis of ordinates (ζ=0) gives the value of the PZC or the IEP.

In an experimental context, the measurements are taken using a zetameter, of which the principle is shown in Figure 6.6.

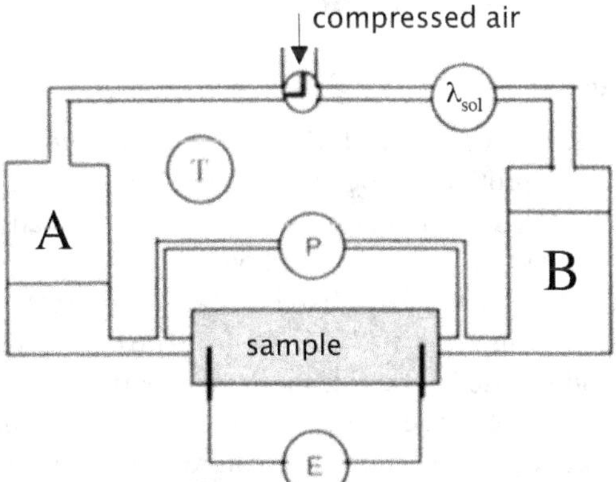

Figure 6.6. *Diagram showing the principle of a zetameter. Based on [CHE 02]*

Figure 6.7 gives an example of how to determine the IEP of a surface of AISI 304 steel, taken from the work by-Boulangé-Petermann *et al.* [BOU 95]. This surface underwent a prior annealing treatment in an atmosphere containing hydrogen (bright annealed).

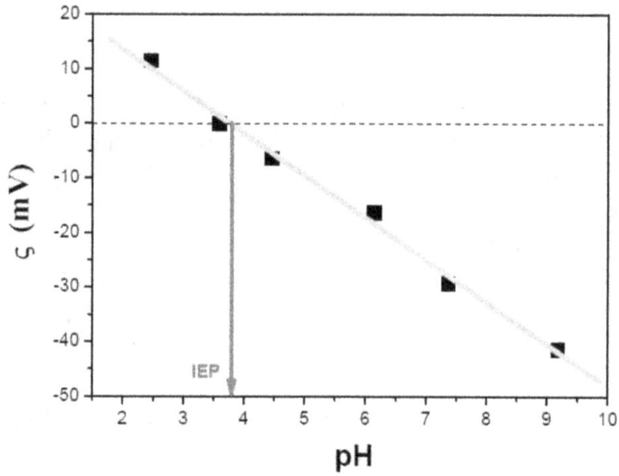

Figure 6.7. *Determining the IEP of a sheet of AISI 304 steel which has undergone a bright annealing treatment. Measurements were taken in a solution of 0.01M NaCl. Taken from Boulangé-Petermann et al. [BOU 95]. Copyright (2007) with permission from Elsevier*

6.2.3. *Scanning probe microscopies*

6.2.3.1. *Atomic force microscopy (AFM)*

Atomic force microscopy (AFM), initially developed by Binnig *et al.* in 1985 [BIN 86] to provide topographic images of surfaces, is based on the measurement of interactions between a nanometric tip and the surface of a sample. The principle of this method is shown in the diagram in Figure 6.8(a). The sample is fixed to a piezoelectric tripod allowing movements XYZ while the tip is fixed to a flexible cantilever. When the sample is brought close to the tip, the repulsive or attractive force between the tip and the surface causes a deflection of the lever d proportional to the force of interaction, which can be monitored using a device coupling the reflection of a laser beam onto the rear side of the cantilever to a four quadrant photodiode detector.

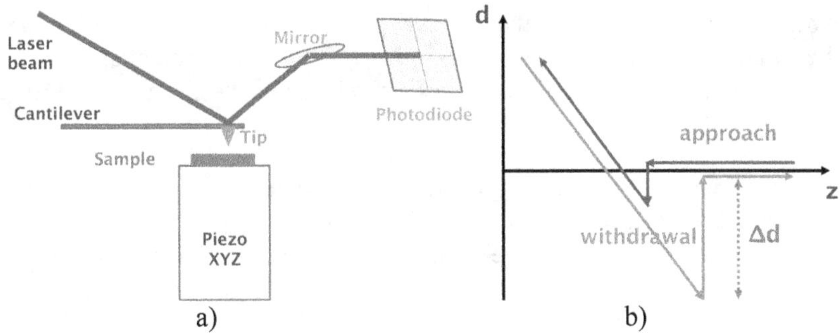

Figure 6.8. *a) Diagram showing the principle of AFM; b) principle of measuring the force of adhesion using force-distance curves*

This device can also be used to measure the force of adhesion between the probe and the surface. In this case, variations in the deflection d of the lever are measured as a function of the distance z between the sample and the tip when the sample is brought close to (approach phase), then moved away (withdrawal phase) from, the tip, while x and y remain fixed. The curve obtained is shown in the diagram in Figure 6.8(b).

As a first approximation, the adhesion force corresponds to the withdrawal force and can be expressed by the relationship:

$$F = k \,.\, \Delta d \qquad [6.4]$$

where k represents the stiffness coefficient of the cantilever.

If k does not vary (using the same measurement device), it is therefore possible to obtain the acid-base properties of the surface by measuring the variations in the force of interaction between the tip and the surface immersed in an electrolyte solution with variable pH. This approach was proposed by Lin *et al.* [LIN 93] and is based on the theory of Derjaguin–Landau–Verwey–Overbeek (DLVO), which provides the expression of the interaction F_{dl} between a sphere of radius R at a distance d from a plane immersed in an aqueous electrolyte solution [ISR 89]:

$$F_{dl} / R = 4 \Pi \varepsilon \varepsilon_0 \phi_1 \phi_2 e^{-d/\lambda} / \lambda \qquad [6.5]$$

where ε and ε_0 represent the relative permittivity of the electrolyte and the vacuum absolute permittivity, Φ_1 and Φ_2 the potentials of the surface of the sphere and the plane respectively, and λ the electrolyte's Debye length.

The sign of F_{dl} therefore originates from the sign of the product $\Phi_1\Phi_2$, and corresponds to repulsive or attractive interactions when Φ_1 and Φ_2 have the same or opposing signs. The sign change therefore takes place at the isoelectric point.

By measuring the forces of interaction between a flat surface and a tip in the electrolytes of variable pH for a fixed distance d, Lin *et al.* [LIN 93] showed that the IEPs of the tip and the flat surface (IEP_1 and IEP_2) could therefore be deduced from the variations of F_{dl} with the pH of the electrolyte (Figure 6.9).

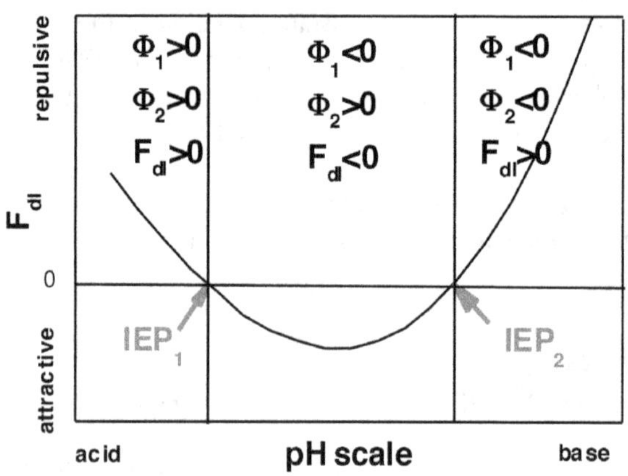

Figure 6.9. *Measurement of the isoelectric points between a tip (IEP_1) and a plane surface (IEP_2). Taken from Lin et al. [LIN 93]. Copyright (1993) with permission from the American Chemical Society.*

A similar approach was taken by Dubois and Joud to measure the isoelectric point of passive films formed on a polished chromium specimen immersed in water, which is obtained for the value of the pH corresponding to the change in the sign of F (IEP = 8.1 in the example given in Figure 6.10) [DUB 00].

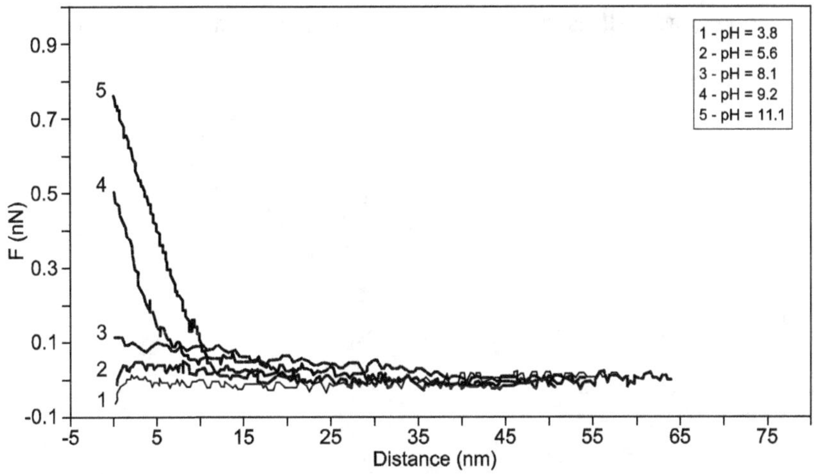

Figure 6.10. *Force-distance curves obtained on oxidized chromium surfaces using a silicon tip covered with* SiO_2*. Taken from Dubois and Joud [DUB 00]*

6.2.3.2. *Interfacial force microscopy (IFM)*

Interfacial force microscopy (IFM) was developed by Joyce and Houston [JOY 91] to overcome the problems of mechanical instability involved in AFM measurements during the approach or withdrawal of the cantilever.

With this device (see diagram in Figure 6.11), the sample is fixed to a piezoelectric tripod allowing movements XYZ. The tip is fixed to a support with an infinite stiffness coefficient (zero compliance), connected to torsion bars. When the sample is brought close to the tip, the attractive or repulsive force between the sample and the tip rotates the rigid support around the axis of the torsion bars. This torsion is compensated by applying a DC voltage to a couple of capacities C_1 and C_2 located under either side of the support of the torsion axis. The value of the voltage necessary for re-establishing the equilibrium therefore provides a direct measurement of the interfacial force. The force-displacement curves can be drawn when the sample is brought close to (approach curve) or withdrawn from (withdrawal curve) the tip. This type of device enables the measurement of forces in the range of 10–50 μN, with a resolution better than 1 nN and a lateral displacement resolution of the order of 1 Å. Consequently, this can be used to detect weak interactions (van der Waals) as well as strong interactions (e.g. acid-base interactions) at a molecular scale.

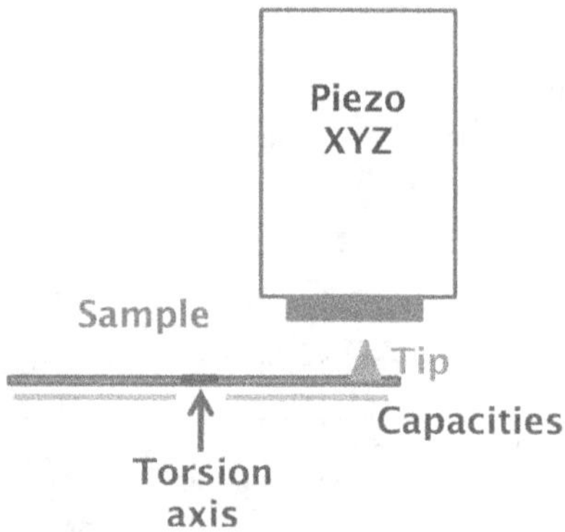

Figure 6.11. *Diagram of the principle of an interfacial force microscope*

The work of adhesion W_A between the tip and the surface of the sample is given by the following relationship:

$$W_A = \frac{-L_c}{\pi CR} \qquad [6.6]$$

where L_c represents the critical charge (or the maximum attractive force upon withdrawal), R the radius of the tip (assimilated to a parabola) and C a constant which depends on the chosen model (C = 3/2 in the Johnson, Kendall and Roberts model (JKR, [JOH 71]) and C = 2 in the Derjaguin, Muller and Toporov model (DMT, [DER 75])).

Figure 6.12 is taken from the work carried out by Houston's group and gives an example of the force (normalized to the diameter of the tip) displacement plot upon approach (unfilled circles) or withdrawal (crosses) between a sample and a tip covered with a gold film and functionalized by a layer of COOH-terminal thiols and a layer of NH_2-terminal thiols, respectively [THO 95]. The work of adhesion deduced from these curves is W_A = 907±83 mJ/m^2 and is attributed by the authors mainly to hydrogen bonds.

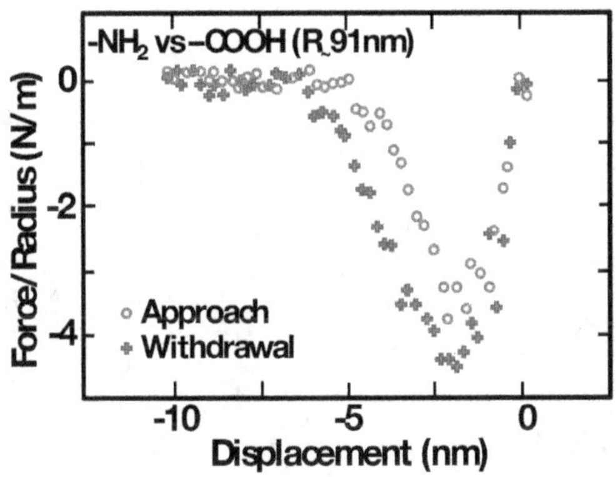

Figure 6.12. *Force (normalized to the radius of the tip) displacement plots obtained between a sample functionalized by thiol alkanes with COOH terminal groups and a tip functionalized by a layer of NH_2-terminal thiols. Taken from [THO 95]. Copyright (1995), with permission of the American Chemical Society*

6.2.4. *Inverse gas chromatography at infinite dilution conditions*

Inverse gas chromatography at infinite dilution conditions (IGC-ID) can be used to evaluate dispersive and polar interactions between a solid material and model chemical molecules, called molecular probes (or adsorbable or solute molecules). It is mainly used to characterize divided solids, but can also be applied in the case of fibers. The solid is inserted into a chromatographic column (Figure 6.13(a)). Using a carrier gas, gaseous probe molecules are injected into the column in very low quantities, in such a way that probe–probe interactions do not need to be taken into account. The concentration of molecules withdrawing from the column is measured as a function of the time each probe takes to cross the column (retention time) and defines the chromatogram (Figure 6.13(b)). Each probe has its own retention time, which depends on its affinity with the solid surface and therefore gives a specific signal in the detector. Molecules with different well-defined properties are used, such as alkanes (in order to determine dispersive reactions) or polar molecules (in order to determine acido-basicity). Thermodynamic laws can be used to calculate the variation in the free adsorption energy between the molecule and the solid, which is connected to the adhesion energy.

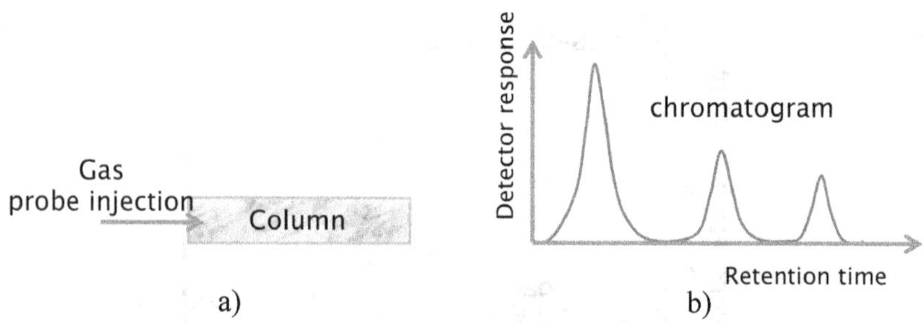

Figure 6.13. *Diagram of the principle of inverse gas chromatography*

The injection of linear alkanes gives access to the dispersive component of the surface energy γ_S^d while the injection of acid or base molecules can be used to calculate the specific components γ_S^{sp} and the acid and base constants K_A and K_D.

In infinite dilution conditions, the net volume of retention V_n (i.e. the volume of carrier gas necessary to allow the probe molecule to cross the column, corrected for the dead volume which is the volume of the column not occupied by the solid) is given by the following relationship:

$$V_n = K.s \tag{6.7}$$

where K is the Henry constant (or the partition constant, which is the partition of probe molecules between the adsorbed and gaseous phases) and s represents the surface accessible to the probe.

The variation in the free enthalpy of adsorption ΔG_a accompanying the passage of the molecule from a gaseous state to an adsorbed state is connected to the net volume of retention V_n by the relationship:

$$-\Delta G_a = RT \ln V_n + C \tag{6.8}$$

where R represents the ideal gas constant, T the temperature in Kelvin and C a constant which depends on the reference state chosen for the adsorbed probe.

Moreover, assuming that the dispersive and specific interactions are additional, ΔG_a can be expressed by the following relationship:

$$\Delta G_a = \Delta G_a^d + \Delta G_a^{sp} \tag{6.9}$$

The comparison in the behavior of non-polar and polar probe molecules can therefore be used to determine the specific component ΔG_a^{sp}, by subtraction of the dispersive component ΔG_a^{d} (obtained with non-polar probes) from the total value ΔG_a (obtained with polar probes). As the specific interactions are dominated by the acid-base interactions, a characterization of the solid's acid-base properties can thus be obtained.

Dorris and Gray showed that the free enthalpy of adsorption during the injection of linear alkane molecules (n-alkanes) varies linearly with the number of carbon atoms n in the probe [DOR 80]. This linear behavior can be generalized to other molecular descriptors (e.g. vapor pressure of the injected probe) and an "*n*-alkanes reference line" can therefore be obtained. In the line showing ΔG_a as a function of this molecular descriptor, the points corresponding to polar probes which interact more strongly with a polar surface are located above the *n*-alkanes reference line. The discrepancy corresponds to the contribution ΔG_a^{sp} of the specific interactions to the free enthalpy variation (Figure 6.14).

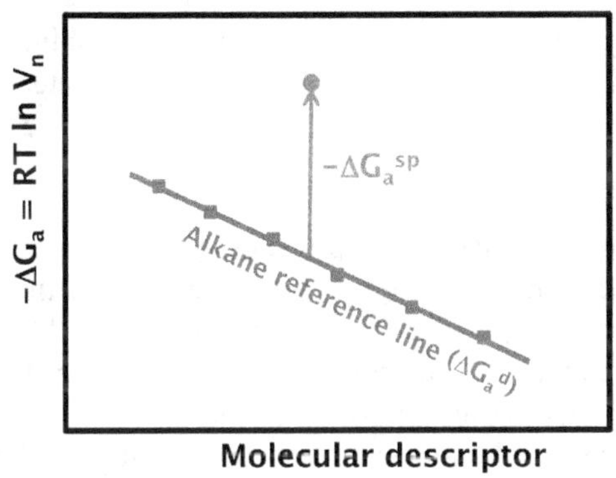

Figure 6.14. *Determining the specific component ΔG_a^{sp} of the free enthalpy of adsorption*

In this representation, the polar probe is assimilated to a molecule which would exchange the same dispersive interactions as an alkane with an identical molecular descriptor. The value of ΔG_a^{sp} therefore depends on the choice of the molecular descriptor, which defines the scale of comparison. Several models of description of this hypothetical molecule have been proposed. These are only briefly described in this book. Any reader interested in this area may consult, for example, the review paper by Brendlé and Papirer [BRE 97].

The first two models proposed are based on the parameters which reflect the intermolecular interactions of the probe molecules at a liquid state. The oldest (1983) was proposed by Saint-Flour and Papirer [SAI 83]. In this model, the probe molecules are characterized by the logarithm of their saturation vapor pressures at the temperature of the column, $\log P^0$. Since the values of P^0 can be easily obtained from the literature, this is the most commonly used method. However, it raises a number of problems for solids with high surface energy. Subsequently, Schultz *et al.* suggested using the product $a(\gamma_l^d)^{1/2}$ where a represents the area of the adsorbed molecule and γ_l^d the dispersive component of the surface energy of the probe in a liquid state, deduced using the expression of Fowkes' work of adhesion [SCH 87]. The main difficulty of this method is determining the value of a, which can depend on the nature of the solid on which the molecule is adsorbed and the temperature. Moreover, the values of γ_l^d are not always accessible at the temperatures used in IGC.

The models developed thereafter are based on the molecular characteristics of the probes. Dong *et al.* and Donnet *et al.* suggested using the product $\alpha_0\sqrt{h\nu}$ where α_0 represents the polarizability of the probe, h Planck's constant and ν the electronic frequency characteristic of the probe [DON 89, DON 91]. Unfortunately, the values of α_0 are seldom available in the literature.

Finally, Brendlé and Papirer proposed using a topological index X_T which takes into account the geometry of the molecule and the local electronic density linked to the presence of heteroatoms [BRE 97]. This parameter represents the number of carbon atoms of a hypothetical hydrocarbon molecule which would interact with the surface in the same way as the molecule it represents. It has the advantage of being directly calculated, for both polar and non-polar molecules, using Wiener indices.

It is important to highlight the fact that the simple approaches presented above are only valid for flat and homogeneous surfaces on a molecular scale. Other models have been developed to take into account the roughness or heterogeneous nature of the majority of solid surfaces, which go beyond the scope of this book. Any reader interested in this area may consult the review paper by Balard *et al.* [BAL 00].

Several approaches have been put forward to quantitatively characterize the acid or base nature of a solid surface. Here, we will only briefly discuss the Gutmann model [GUT 78], which is most commonly used in IGC-ID.

Any reader interested in this area may refer to the review paper by Mukhopadhyay and Schreiber [MUK 95].

In the Gutmann model [GUT 78], the Lewis acido-basicity of the solid surface is characterized by two parameters K_A and K_D, representing respectively the solid's acid and base constants, which are linked to the donor (DN) and acceptor (AN*) numbers of the probe by the following relationship:

$$\Delta H_a^{sp} = K_A\, DN + K_D\, AN^* \qquad [6.10]$$

where ΔH_a^{sp} represents the specific enthalpy of adsorption, linked to the free enthalpy ΔG_a^{sp} and the specific entropy ΔS_a^{sp} by the following relationship:

$$\Delta G_a^{sp} = \Delta H_a^{sp} - T\,\Delta S_a^{sp} \qquad [6.11]$$

In practice, the value of ΔH_a^{sp} is obtained by measuring the slope of the linear plot $\Delta G_a^{sp}/T = f(1/T)$. The donor DN and acceptor AN* numbers are empirically determined by studying the interactions of the probe with a reference base (triethylphosphine, Et_3PO) and a reference acid (antimony pentafluoride, $SbCl_5$).

DN and AN* are easily obtained in the literature. K_A and K_D are graphically obtained using the linear plot $\Delta H_a^{sp}/AN^* = f(DN/AN^*)$. K_A corresponds to the slope of the plot and K_D to the ordinate at the origin.

Application example

The following application example is taken from the work of Reddi Rani *et al.* [RED 11].

Cellulose acetate butyrate (CAB) is a thermoplastic in the cellulosic family, very commonly used in the sectors of blister packaging, tool handles, eyewear and the inner trim of cars.

Figure 6.15(a) shows the values of $-\Delta G_a$ obtained at 333.15 K with *n*-alkanes non-polar probes ($n = 5$–8) and with polar probes: dichloromethane (DCM), trichloromethane (TCM), acetone (AC), ethyl acetate (EA) and tetrahydrofuran (THF). The model of the molecular descriptor used for the scale comparison is that developed by Schultz *et al.* [SCH 87]. The values of ΔG_a^{sp} are deduced as indicated for DCM.

Similar measurements carried out by steps of 10 K in the temperature range 333.15–393.15 K helped to determine ΔH_a^{sp} using the slope of the plot $\Delta G_a^{sp}/T = f(1/T)$. The values of the acid constant K_A=0.126 and base constant K_D=1.109 deduced from Figure 6.15(b) show that the surface has a base character, which was attributed by the authors to the ester groups of the CAB.

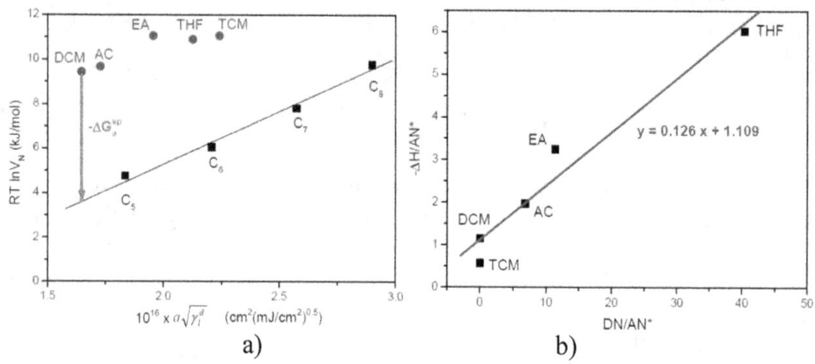

Figure 6.15. *a) Variations in the free enthalpy of adsorption ΔG_a on a Cellulose Acetate Butyrate (CAB) surface as a function of the probes used: non-polar (n-alkanes with n = 5 to 8) or polar: dichloromethane (DCM), trichloromethane (TCM), acetone (AC), ethyl acetate (EA) and tetrahydrofuran (THF). The value of the specific component of the free energy ΔG_a^{sp} is indicated in the case of the DCM. b) The linear plot $\Delta H_a^{sp}/AN = f(DN/AN)$ gives the values of K_A (slope) and K_D (ordinate at the origin). Taken from [RED 11], copyright (2011) with permission from Wiley*

6.2.5. *X-ray photoelectron spectroscopy (XPS)*

6.2.5.1. *The principle of XPS*

X-ray photoelectron spectroscopy (XPS), initially known as electron spectroscopy for chemical analysis (ESCA), is based on the photoelectric effect. This technique was developed from the 1960s, when Siegbahn and his collaborators [SIE 67] identified its capacity of giving information on the chemical state of the samples.

The interaction of a photon X of energy $h\nu$ with an atom or molecule leads to the ionization of this atom or molecule and the emission, from the energy level E_1, of a photoelectron of kinetic energy $E_K = h\nu - E_1$ (see Figure 6.16). The hole created by the photon is filled by an electron from a higher level, with energy E_2. The relaxation of the excess energy (E_1-E_2) available can be done in two ways: either a photon is emitted with the characteristic energy hν' = E_1-E_2 (phenomenon of fluorescence), or an Auger electron is ejected from an external layer (of energy E_3) with the kinetic energy $E_A = E_1 - E_2 - E_3$* (*takes into account the presence of a hole in the emitting atom).

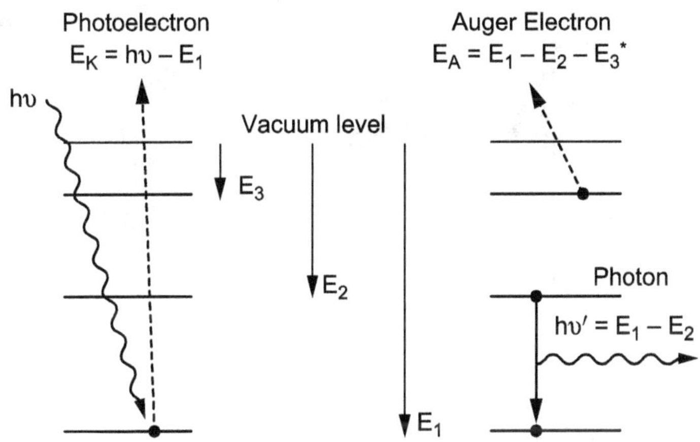

Figure 6.16. *Principle of photoemission*

The kinetic energy of a photoelectron depends solely on the energy of the incident photon hν and on the binding energy of the ejected electron E_1 and is therefore characteristic of the emitting element.

The photoelectric transitions are described using the classification of the level where the photoelectron comes from: 1s, 2s, $2p_{1/2}$, $2p_{3/2}$, etc.

X-ray induced photoelectron spectroscopy (or X-ray photoelectron spectroscopy, XPS) is based on the measurement of the energy distribution of the electrons emitted by a surface bombarded by an X-ray beam of energy of the order of keV. Therefore, it detects both the photoelectrons and Auger electrons. The energy of the photoelectron, referenced to the Fermi level of the sample, can be expressed in terms of kinetic energy E_K or binding energy E_B, with these two values being linked by the following relationship:

$$E_B = h\nu - E_K - \Phi_{sp} \qquad [6.12]$$

where Φ_{sp} represents the work function of the spectrometer.

A typical example of an XPS spectrum is shown in Figure 6.17 (XPS spectrum of copper). The XPS peaks characteristic of the most intense core levels (Cu2p and Cu3s) are indexed and appear together with the spectrum of the valence band in the region 0–30 eV and the Auger transitions Cu LMM.

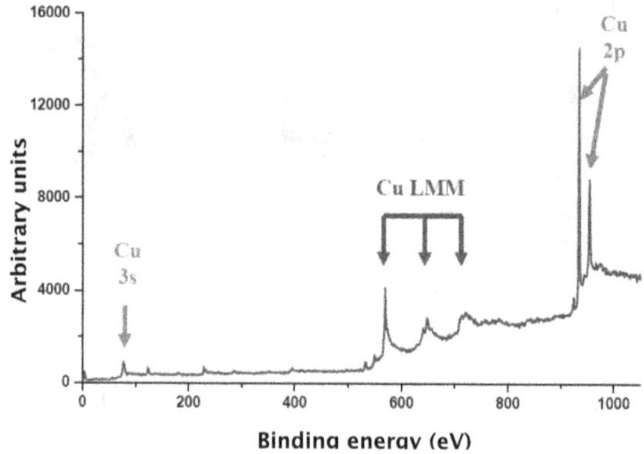

Figure 6.17. *XPS spectrum of copper*

The analyzed depth is directly dependent on the mean free path of the electrons emitted, and thus on their kinetic energy. In XPS, this depth is typically around a few nm. It should be noted that it is possible to reduce this depth (and therefore increase the sensitivity of the surface) by increasing the angle between the analyzer and the normal to the surface (angular analysis).

6.2.5.2. *Correlations between electronic structure and acid-base properties*

The measurement of acid-base properties of a solid surface using XPS is based on an analogy between the band structure of a solid and the molecular orbitals of a molecule.

The first description of the acid-base interactions in terms of electronic structure was given by Lewis, who considered an acid as an electron acceptor and a base as a donor [LEW 23]. Mulliken [MUL 51], and later Klopman [KLO 68] expanded on Lewis' concept by describing the chemical bond between two molecules by a sharing of the electrons from the highest occupied molecular orbital (HOMO) of the base with the lowest unoccupied molecular orbital (LUMO) of the acid, leading to the formation of bonding and anti-bonding molecular orbitals of the adduct (see Figure 6.18).

Figure 6.18. *Diagram of the principle of an acid-base reaction by sharing an electron pair*

In this model, the acid or base nature of a species is therefore solely dependent on the relative positions of its HOMO and LUMO levels compared to those of other species. When the gap between the LUMO and HOMO is too great, the interaction between orbitals is no longer possible and a complete electronic transfer leads to the redox phenomenon [SAT 89].

Although this model does not take into account the energy differences between HOMO and LUMO within the various species nor the perturbation due to orbital overlapping, it has been extended to solid surfaces by assimilating the HOMO and LUMO orbitals respectively with the valence band and the conduction band at 0K. The Lewis acidity of the surface can therefore be deduced from the binding energy E_b at the core levels observed with XPS.

An initial approach proposed by Mullins and Averbach [MUL 88a, MUL 88b, MUL 88c], linked the acid-base nature of an oxide surface to the binding energy of the core levels of the oxide constituents. This approach, adopted by Casamassima *et al.* [CAS 91a, CAS 91b], is based on the following expression:

$$E_B = -\varepsilon_F + E_R + E_{F0} \qquad [6.13]$$

where $-\varepsilon_F$ represents the energy of the Hartree–Fock orbital corresponding to the electron emitted in its fundamental state, E_R the relaxation energy corresponding to reorganization after photoemission and E_{F0} the energy of the Fermi level referred to the vacuum level.

When the chemical environment of the investigated atom varies, the variation induced in the density of valence electrons modifies the screening of the core levels and leads to a displacement ΔE_B of the binding energy which can be written as follows:

$$\Delta E_B = -\Delta\varepsilon_F + \Delta E_R + \Delta\, E_{F0} \qquad [6.14]$$

The last two terms are generally neglected and the variations observed are therefore attributed to the sole modification of the energy of the core level due to the variation in the chemical environment (chemical shift).

Mullins and Averbach proposed that, for similar oxides, the local electronic density around ions is similar and the variation in the energy of electronic levels is therefore negligible. The variations ΔE_B observed in XPS are thus attributed to a displacement of the Fermi level relative to the bottom of the valence band and are identical for all the same photopeaks. In this model, the higher the energy in the Fermi level E_F relative to the valence band, the more acidic the surface is. This interpretation is based on the agreement observed between the variations in the binding energy deduced from the XPS spectra and the PZC measurement obtained for Mg-Al-Si oxide powders with different cation contents (see Figure 6.19).

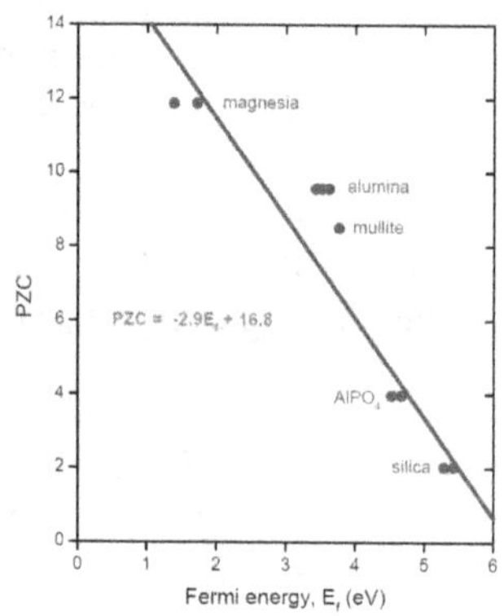

Figure 6.19. *Relationship between the PZC and variation in the Fermi level deduced from measurements of binding energy in XPS. Taken from [MUL 88b], copyright (1988) with permission from Elsevier*

Nevertheless, the interpretation given by Mullins and Averbach attributing the variation in binding energy to the variations in the Fermi level appears questionable. In fact, in the rigid band model used by Mullins and Averbach and retaining the hypothesis of equivalent electronic structures, the increase in the Fermi level (relative to the bottom of the valence band) is the direct consequence of the increase in the population of the conduction band (or of localized states in the oxide gap, as proposed by Baroux and Gorse [BAR 94]), i.e. an increase in the density of electronic donor states. It should therefore correspond to an increase in the Lewis basicity of the surface rather than in its acidity as proposed by Mullins and Averbach. Indeed, previous work carried out by Butler and Ginley [BUT 78] and by Labib and Williams [LAB 84] connect the intrinsic increase in the Fermi level to an increase in basicity.

This apparent contradiction comes from the fact that, during the acquisition of an XPS spectrum, the spectrometer/analyzer set is electrically connected to the ultra-high vacuum chamber, i.e. to the ground. The Fermi level of the sample analyzed is therefore not its intrinsic level but results from a balance with the spectrometer Fermi level. In fact, Debontridder [DEB 01] showed that the displacements measured after correction of the charge effect are not the direct measurement of the displacement of the intrinsic Fermi level, but of the difference between the variation in the energy of the core levels and that of the intrinsic Fermi level relative to the vacuum level. According to Debontridder's hypothesis, the increase in the energy of the core levels is therefore correlated to a decrease in the energy of the Fermi level relative to the vacuum level, and consequently to an increase in the acidity of the surface, as predicted by Lewis' theory.

Even though the interpretations of the shifts in the core level peaks differ, the correlation between the binding energies and acid-base properties has been observed by a number of authors. The measurement of the binding energies can also be used to characterize the acid-base properties of polymer surfaces and any reader interested in this area may refer to the review paper by Chehimi *et al.* [CHE 00].

The comparison of the energy levels measured for various surfaces can therefore be used as a classification of the acid-base nature of these surfaces, as shown by the following example, taken from the work carried out by Debontridder [DEB 01].

Aluminum coupons with very high purity (99.999%) were mirror polished, then underwent one of the following treatments:

– thermal: annealed at 300°C for 24 h (TT300) or annealed at 500°C for 24 h (TT500);

– anodizations in a solution of pH = 8.2 containing 0.1M H_3BO_3 +0.01M $Na_2B_4O_7$. Two types of anodizations were carried out: at 25°C on a polished and degreased surface (AN25) or at 95°C after flash heat treatment at 600°C for 10 min (AN95);

– hydrothermal (boehmitage), by immersion for 10 min in boiling distilled water (16 MΩ.cm) on a polished and degreased surface (BOEH).

Figure 6.20 represents the variations in the binding energy of the component of the O1s peak corresponding to oxidized aluminum as a function of the binding energy of the component of the Al2p peak corresponding to oxidized aluminum ($Al2p^{ox}$) after correction of the charge effect. This figure shows a linear correlation between O1s and $Al2p^{ox}$. A line (shown by the solid line on the graph) of slope 1.2 is obtained, close to the theoretical line (dotted line) obtained with a regression of slope 1, thus enabling the surfaces to be classified in terms of increasing acidity:

$$BOEH < TT300 < TT\ 500 < AN25 \leq AN95$$

The BOEH treatment produced the most basic surface and the anodic oxides (AN95 and AN25) are the most acidic.

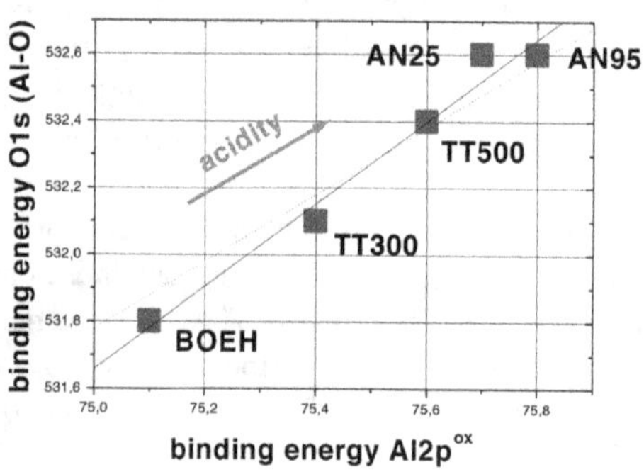

Figure 6.20. *Variations in the binding energies of the components of peaks O1s and Al2p corresponding to the Al-O bonds of pure aluminum surfaces after undergoing different treatments. Taken from [DEB 01]*

6.2.5.3. *Other uses of XPS*

The acid-base nature of a surface can also be determined using exchange reactions between the surface sites and one or more species reacting specifically with certain sites. Since XPS is in essence a surface analysis technique, it can therefore be used to identify and quantify these exchange reactions. This method was initially proposed by Simmons and Beard to determine the strength of the acid character of the hydroxyls localized at the surface of hydrated iron samples [SIM 87]. The method is based on determining equilibrium constants K_1 and K_2 of the following reactions in aqueous environments:

$$-FeOH_2^+ \leftrightarrow -FeOH + H^+ \text{ (constant } K_1\text{)} \qquad [6.15]$$

$$-FeOH \leftrightarrow -FeO^- + H^+ \text{ (constant } K_2) \qquad [6.16]$$

The values of K_1 (K_2) can be deduced from the variations in the fraction θ of hydroxyls for which hydrogen was exchanged with an anion (cation) as a function of the pH of the solution. For instance, for an exchange with a potassium ion, the following relationship can be used:

$$pH = pK_2 - log\frac{\theta}{1-\theta} \qquad [6.17]$$

The value of pK_2 corresponds to the pH of the solution for which $\theta = 0.5$. The measurements carried out by Simmons and Beard led to values of $pK_1=8.5$ and $pK_2=11.5$ allowing the value of the isoelectric point IEP to be deduced using the following relationship:

$$IEP = \frac{1}{2}(pK_1 + pK_2) \qquad [6.18]$$

This method has been successfully used by other authors to estimate the isoelectric points of metals covered with oxide layers [WAT 91, KUR 92, KUR 93].

6.2.6. *Other methods*

Although the methods described above are the most commonly used, several other experimental methods can be implemented to classify surfaces in terms of acido-basicity. In particular, a number of methods derived from electrochemistry have been suggested to identify the acid or base properties of passive films formed on metallic substrates.

Passive films can be considered as wide gap semiconductors. The electroneutrality of a passive film immersed in an electrolyte can therefore be written as follows:

$$\sigma_m + \sigma_{SC} + \sigma_0 + \sigma_a + \sigma_d = 0 \qquad [6.19]$$

where σ_m represents the charge at the metal–passive film interface, σ_{SC} the space charge in the passive film caused by the transfer of charge at the electrolyte-film interface, σ_0 the surface charge of the film, σ_a the charge corresponding to specific adsorption on the surface of the passive film and σ_d the charge of the diffuse layer.

Several approaches have been proposed. Casamassima *et al.* showed that the electrochemical potential in an open circuit increases with the acidity of the surfaces [CAS 91b]. Lopez *et al.* suggested using the space charge σ_{SC} as an acid-base parameter [LOP 98]. It has the same sign as the band bending $e\Delta\Phi_{SC}$ within the film and is connected to it by the following expression:

$$\left|\sigma_{SC}\right| = \sqrt{2\varepsilon_0\,\varepsilon_{SC}\,Ne.}\left|\Delta\phi_{sc}\right| \qquad [6.20]$$

where ε_0 represents the dielectric constant of vacuum, ε_{SC} the dielectric constant of the passive film, N the density of the charge carriers in the passive film and e the absolute value of the electron elementary charge.

The values of $e\Delta\Phi_{SC}$ can be obtained using photoelectrochemistry by measuring the variations in the quantum efficiency as a function of the energy of incident photons on a sample kept at rest potential. The more negative (or positive) the charge (or the band bending), the more basic (or acid) the surface is.

In fact, for surfaces showing similar flat band potentials (e.g. films formed on the same family of alloys), the two approaches described above give the same trend. This can be explained by the fact that the band bending is the difference between the flat band potential and the rest potential. Moreover, equation [6.20] shows that the space charge depends on the rest potential, the flat band potential and the density of charge carriers. Methods such as electrochemical impedance spectroscopy or chronoamperometry, which can measure the density of charge carriers, can thus also provide information about acid-base properties [MER 99].

6.3. Local methods

The approaches described in section 6.2 provide a general description of the acid-base properties of a surface, in terms of surface charge, point of zero charge or isoelectric point. However, solid surfaces have both acid and base sites and their overall acido-basicity is a consequence of the nature, the distribution and the strength of these various sites. A more local description to determine these parameters is therefore often necessary in various applications. This type of local description can be obtained directly by using a technique which is sensitive to the acid-base nature of the sites on the surface or indirectly by monitoring the adsorption of surface probe molecules to be characterized. It is important to highlight that, in the second case, the characteristics of an acid (or base) site can depend on the probe molecule used. Probe molecules are usually small in size, so as to probe all sites on the surface without being sterically hindered. For instance, ammonia, pyridine, carbon monoxide, acetic acid, carbon dioxide or sulfur dioxide are all commonly used. However, it should be noted that larger molecules can also be used, either to select a specific type of site (e.g. lutidine can be used to only probe Brønsted acid sites, since the two methyl groups on both sides of the nitrogen atom prevent access to cations), or to model a specific reactive function. Probe molecules are therefore chosen for their ability to produce a significant and easily measureable signal of a given mode of adsorption. The most commonly used experimental techniques to characterize surface acid or base sites are infrared spectroscopy and photoelectron spectroscopy.

6.3.1. *Infrared spectroscopy*

6.3.1.1. *Principle of infrared spectroscopy*

Within a molecule, atoms are permanently vibrating around a mean position and bond lengths and angles are constantly changing, thus generating an oscillating electric field. When the frequency of this electric field is equal to the frequency of the electrical component of the incident infrared radiation, absorption occurs. Therefore, a molecule can only absorb a radiation whose frequency is identical to the vibration frequency which generates a variation in its dipole moment. The spectroscopic domain used to measure molecular vibrations is infrared, ranging from 4,000 cm^{-1} to 400 cm^{-1} (corresponding to wavelengths of 2.5–25 μm).

In the spectrum measuring the variation in the radiation intensity as a function of its frequency (or wavenumber), each absorption gives rise to a peak characteristic of a given bond thus enabling the molecular group to be identified, since its wavenumber is almost independent from the molecule to which it belongs. It is therefore possible to identify chemical groups (characterized by their wavenumber) and to evaluate their concentration (using the size of the absorption peak). Table 6.1 provides several vibration frequency values characteristic of molecular groups. It should be noted that the values corresponding to an absorbed molecule on a solid surface are usually slightly different from those corresponding to the same molecule in a gaseous state.

Type of bond	Group	Wavenumbers (cm^{-1})
Elongations		
C-H	C_{sp3}-H	2800 – 3000
	C_{sp2}-H	3000 – 3100
	C_{sp}-H	3300
C-C	C-C	1150 – 1250
	C=C	1600 – 1670
	C≡C	2100 – 2260
C-N	C-N	1030 – 1230
	C=N	1640 – 1690
	C≡N	2210 -2260
C-O	C-O	1020 – 1275
	C=O	1650 – 1800
N-H	RNH_2, R_2NH	3400 – 3500
	RNH_3^+, $R_2NH_2^+$, R_3NH^+	2250 – 3000
O-H	ROH	3610 – 3640 (free) 3200 – 3400 (H bonds)

Table 6.1. *Characteristic vibration frequencies of some molecular groups*

Several variations of infrared spectroscopy allow the phenomena of adsorption to be studied and therefore enable the acid-base properties of solid surfaces to be characterized. However, most methods can only be used to study particles. This book only presents infrared reflection absorption spectroscopy (IRAS), which can be used to study solid surfaces with a low specific surface. Any reader interested in the other techniques may refer to specialized works.

6.3.1.2. *Infrared reflection absorption spectroscopy (IRAS)*

Initially developed by Francis and Ellison [FRA 59], IRAS is based on the study of the specular reflection of an infrared beam on a reflective surface. It is mainly used to study thin films or molecular systems deposited on flat metal surfaces.

The principle is illustrated in Figure 6.21. The infrared beam impinges the surface with an incidence angle Φ from the surface normal. The incident electric field can be decomposed into a perpendicular component (polarized s) and a parallel component (polarized p) to the plane of incidence. The polarized component s is only made up of one component parallel to the surface (denoted as S on the diagram). The polarized component p (denoted as P) can be decomposed into a component parallel to the surface (P_t) and one perpendicular to the surface (P_n).

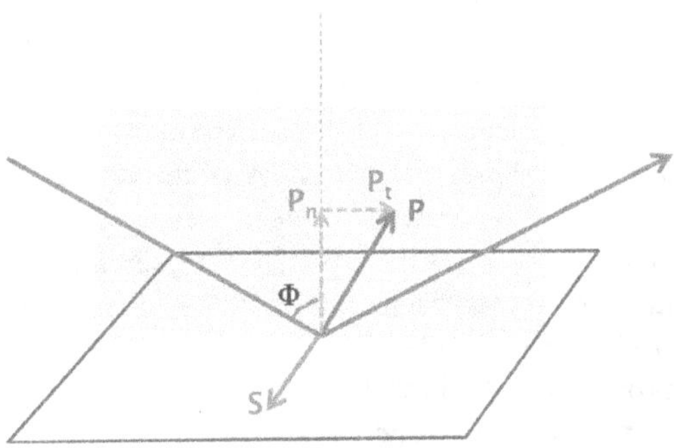

Figure 6.21. *Reflection of an electric field at the surface of a metal*

In the case of a metal substrate, it can be shown that the reflection on the surface produces a destructive interference for the component parallel to the surface whereas a constructive interference is obtained for the component perpendicular to the surface. The absence of an electric field parallel to the surface which can interact with the dipole moment of a vibration mode parallel to the surface explains the surface selection rule, strict in the case of metals, whereby only vibration modes perpendicular to the surface can be detected in infrared spectroscopy. Moreover, a grazing incidence (φ around 80–90°) helps to optimize the signal.

The IRAS spectrum corresponds to the variation of reflectance associated with the adsorbed film compared with the bare substrate, which serves as a reference:

$$\Delta R/R_0 = (R_0\text{-}R)/R_0 \qquad [6.21]$$

where R represents the reflectance associated with the adsorbed film and R_0 the reflectance of the bare substrate, with $R_i=r_i^2$, given that r_i is the reflection coefficient.

The sensitivity of this method is linked to the variations of $\Delta R/R_0$, which are usually low (10^{-1} to 10^{-3}) but can be sufficient to detect monolayer fractions for molecules with significant dipole moment variations. A considerable improvement in sensitivity can be obtained by rapidly modulating the polarization of the incident beam between polarizations p and s using a photoelastic modulator (polarization modulation – infrared reflection absorption spectroscopy (PM-IRAS)). Due to the surface selection rules, this method can also be used to determine the orientation of molecular groups present on the surface.

Information on the nature and strength of reactive sites for a given compound are usually deduced from the vibration frequency of specific molecular groups. For instance, the frequency of the elongation of isolated OH groups on an oxide surface increases (or decreases) when the tendency toward proton dissociation decreases (or increases), corresponding to an increase in the basicity (or acidity) of these groups. Nevertheless, most characterizations are carried out using probe molecules and the acid-base characteristics therefore depend on the probe used. As a result, it is not possible to establish a universal scale of acidity (or basicity) for solid surfaces.

Moreover, the quantitative use of this technique is based on the hypothesis that $\Delta R/R_0$ varies linearly with the concentration of the chemical group responsible for the adsorption. In practice, it is often necessary to carry out a calibration process using another technique.

Application example

The following application example is taken from the work carried out by Freund's group [BOS 13].

A two-dimensional (2D) alumino-silicate film deposited onto a (0001) surface of ruthenium was partially oxidized using D_2O. This film presents bridged groups Si-OD-Al at the surface and can be considered as a zeolite model. These bridged groups generally have high Brønsted acidity and can react with weak bases (such as carbon monoxide or ethylene) as well as strong bases (such as ammonia or pyridine). Figure 6.22 shows the IRAS spectrum obtained after adsorption of CO, divided by the spectrum recorded before exposure to CO. The positive (negative) peaks represent the disappearance (appearance) of the corresponding molecular groups. Figure 6.22 shows a modification of the vibrations of the elongation of the bridged OD groups, which appear at 2,652 cm^{-1} on a clean surface and are shifted to 2,409 cm^{-1} after adsorption of CO. This red shift of the OD groups is attributed to the formation of a CO…D complex between the molecule and the bridged OD groups. The amplitude of the shift observed (243 cm^{-1}, comparable to the higher values observable for zeolites) shows that the OD sites of the 2D alumino-silicate film are strong acids. Moreover, the peak characteristic of CO vibrations appears at 2,183 cm^{-1}, which corresponds to a blue shift of 40 cm^{-1} compared with the CO molecule in a gaseous phase, further evidencing a strong interaction of this molecule with the surface. Finally, since IRAS only allows dipole moments with a component perpendicular to the surface to be observed, the appearance of a peak characteristic of CO vibrations shows that this molecule is not adsorbed parallel to the surface. The upper band in Figure 6.22 shows the values obtained by DFT calculation and confirms the previous conclusions.

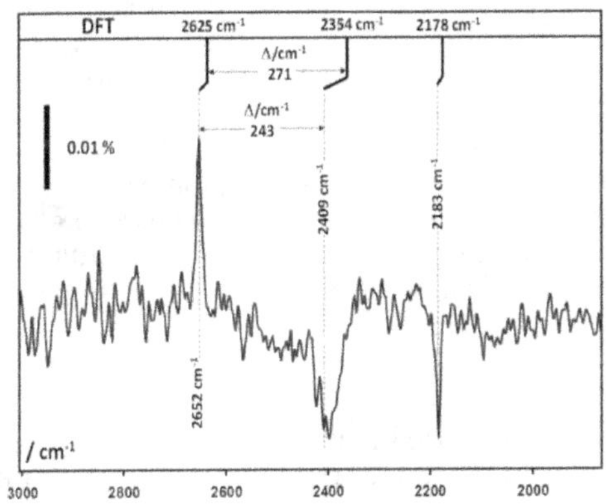

Figure 6.22. *IRAS spectrum obtained after adsorption of CO on a two-dimensional zeolite model deposited onto a Ru(0001) surface. Taken from [BOS 13], copyright (2013) with permission of the American Chemical Society*

6.3.2. *X-ray photoelectron spectroscopy (XPS)*

The principle of x-ray photoelectron spectroscopy was presented in section 6.2.6.1. Section 6.2.6.2 also showed that, for materials with similar chemical compositions, it is possible to classify surfaces in terms of acido-basicity using variations in the binding energy of characteristic peaks in the XPS spectrum. As for the case of infrared spectroscopy, probe molecules can be used to obtain more local information on the nature and density of active sites.

This approach is based on the fact that when a probe molecule reacts with surface sites of different natures, the differences between the probe-reactive site interactions will give rise to a difference in the binding energy of the probe molecule reactive group (see diagram in Figure 6.23(a)). Initially used to probe the nature of the reactive sites of dispersed solids such as zeolites [DEF 78], this approach was then extended to flat solid surfaces.

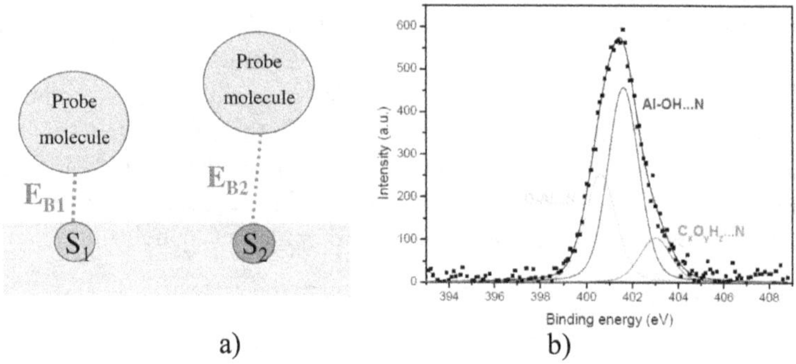

Figure 6.23. *a) Diagram of the principle of interaction between a probe molecule and two surface sites of different natures. b) Components of the N1s peak obtained after the immersion of a 1050 aluminum alloy plate in a xylene solution containing 0.1 mol L^{-1} DAE. See text. Taken from [MER 08], copyright (2008) with permission from Elsevier. For a color version of the figure, see www.iste.co.uk/joud/surfaces.zip*

Most probe molecules used react via groups containing carbon, oxygen, nitrogen, sulfur or fluorine atoms. The binding energies of the core electrons of these atoms can vary from several tenths of electron volts to several electron volts depending on the type of interaction in which they are involved. Using XPS to monitor the adsorption of a probe molecule, with a known acid or base nature, helps to show the acid-base properties of surface sites with which it reacts. For instance, Figure 6.23(b), taken from the work carried out by Mercier *et al.* [MER 08], shows the N 1s peak obtained after immersion of a 1050 aluminum alloy plate in a xylene solution containing 0.1 mol L^{-1} 1,2-diaminoethane (DAE) a strong base. The presence on the aluminum surface of reactive sites of increasing acidity is evidenced by the appearance of three components in the N1s peak, corresponding to three types of interactions. The peak at 400.5 eV can be attributed to O-Al…N donor-acceptor (Lewis-like) interactions between the nitrogen lone pair and the aluminum cations (Lewis acids). The peak at 401.6 eV corresponds to the protonation of the amine termination of the molecule by the surface hydroxyls (Brønsted acids), to form Al-OH…N complexes. Finally, the peak at 402.6 eV is related to the protonation of the amine termination by carbon hydroxyls coming from carbonaceous contamination, which are also

Brønsted acids and form $C_xO_yH_z$...N complexes. It should be noted that the higher binding energy observed for interactions with contamination hydroxyls shows that these hydroxyls have stronger Brønsted acidity than the surface hydroxyls. The relative intensity of the three components observed can also be used to evaluate the relative concentration of each type of reactive site. Finally, the evolution of these relative intensities with time also gives information on the relative stability of these interactions. These last two points are illustrated in Figure 6.24, also taken from the work carried out by Mercier *et al.* [MER 08]. When the XPS analysis is carried out immediately after the DAE adsorption, the interactions with Brønsted acid sites (surface hydroxyls and contamination hydroxyls) represent almost 75% of all of the interactions. However, after 7 days of aging in ultra-high vacuum, these interactions represent less than 60% of the total, therefore showing the greater stability of the interactions with Lewis acid sites (metal cations).

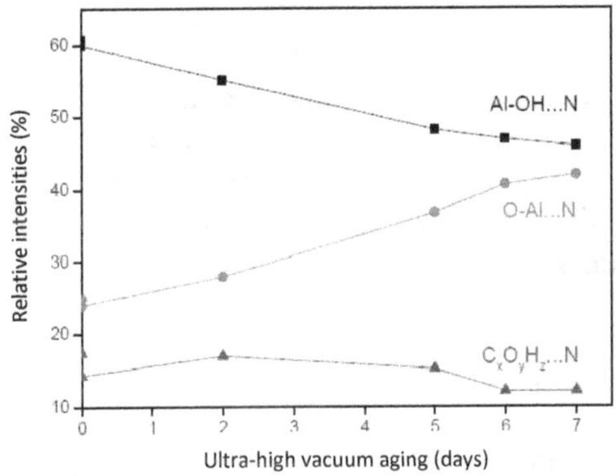

Figure 6.24. *Effect of aging in ultra-high vacuum on the relative intensities of the various components of the N1s peak obtained after the immersion of 1050 aluminum alloy plates in a xylene solution containing 0.1 mol L^{-1} DAE. Taken from [MER 08], copyright (2008) with permission from Elsevier*

6.3.3. Other methods

The popularity of infrared spectroscopy and photoemission spectroscopy to evaluate the acido-basicity of a surface is due to the fact that these techniques are available in many laboratories. However, similar information can be obtained using a number of other experimental methods. In fact, any technique able to assess the adsorption of probe molecules and to characterize their interactions with the surface can, in principle, be used to determine the local acid-base character of surface sites.

For instance, High Resolution Electron Energy Loss Spectroscopy (HREELS) can provide information similar to that obtained by IRAS. It is based on the measurement of energy losses of several tenths to several hundredths of meV occurring when a very low energy electron beam (a few eV) is interacting with a molecule adsorbed onto a surface. The spectrum obtained is a vibrational signature of the adsorbed molecule and helps to determine the nature of the adsorbed molecules, the site and the adsorption geometry, and in some cases, to measure the binding energy. It has the advantage of being more sensitive than infrared spectroscopy, showing modes which are inactive in infrared and intrinsically being a surface analysis method. However, its weaker resolution added to the requirement of working in ultra-high vacuum limit its use.

6.4. Application examples

Although the concept of acido-basicity was initially developed to describe phenomena in aqueous environments, the acid-base properties of solid surfaces partially affect the behavior of a solid with its surroundings and is therefore relevant in a number of application domains such as: adhesion, bio-adhesion, catalysis, corrosion, soil sciences, etc. The following sections provide some examples taken from various domains.

6.4.1. *Bonding ability of aluminum sheets*

Adhesion is a complex phenomenon involving several types of interfacial forces. The role of acid-base interactions in adhesion phenomena, initially highlighted by Fowkes in the 1960s [FOW 64], has since been subject to numerous theoretical and experimental works. The example below is taken from the work carried out by Lopez [LOP 97].

The objective of Lopez's work was to correlate the acid-base properties of surfaces of aluminum alloy plates with their ability to bond with an organic coating.

Due to the variety of the industrial applications of aluminum (food, automobile, aeronautic, electronic, construction sectors, etc.) Lopez used several grades of alloys in his study. Here, we will only present the results relating to grades AA 1198 and AA 5182, often used in the food industry and whose chemical composition (in mass%) is shown in Table 6.2.

	Si	Fe	Cu	Mn	Mg	Cr	Zn	Ti	Al
AA 1198	0.01	0.006	0.006	0.006	–	–	0.01	0.006	>99.98
AA 5182	0.2	0.35	0.15	0.2-0.5	4-5	0.1	0.25	0.1	rest

Table 6.2. *Chemical composition (in mass%) of AA 1198 and AA 5182 alloys*

Grade AA 1198 samples underwent chemical brightening, followed by various surface treatments:

– T1: sodium hydroxide degreasing (15s immersion in 1M NaOH deionized water solution at room temperature, followed by rinsing in deionized water);

– T2: alkaline detergent degreasing (10s immersion in 35g L^{-1} Oakite™ in deionized water at 65°C, followed by rinsing in deionized water);

– T4: T1 followed by hydrothermal oxidation (10 min immersion in boiling deionized water);

– T5: T1 followed by thermal treatment (heating for 4 hours in a muffle furnace at 420°C);

– T7: T1 followed by phosphoric acid anodization (40g L^{-1} H_3PO_4 in deionized water; 20 V DC against a 1199 aluminum foil (Al > 99.99) during 25 min at room temperature, followed by rinsing in deionized water).

Samples of grade AA 5182, in an as-rolled condition, underwent a specific industrial annealing process.

Figure 6.25 shows the correlation between measurements of band bending $|\Delta\Phi_{SC}|$ obtained using the spectroelectochemical method described in section 6.2.7 and the peel force deduced from peeling tests using the acidic SCOTCH 3M 616 adhesive tape applied to treated sheets. It should be noted that all the surfaces gave negative band bending and therefore have a base nature. The increasing abscissae correspond to increasing basicity. Figure 6.25 therefore shows that the practical adhesion of the sheet-adhesive assembly increases with the basicity of the aluminum sheet surfaces.

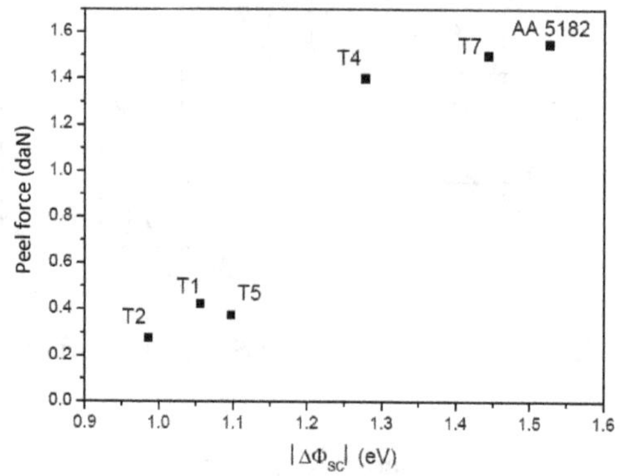

Figure 6.25. *Correlation between the band bending of AA 1198 aluminum sheets (after various surfaces treatments) and AA 5182 and the peel force of an adhesive tape. T1: sodium hydroxide degreasing; T2: alkaline detergent degreasing; T4: hydrothermal treatment; T5: thermal treatment; T7: anodization. Taken from [LOP 97]*

6.4.2. *Mechanism of formation of the interphase in metal–polymer joints*

Epoxy-amine–metal joints are commonly used, especially in aeronautics and the automotive industry.

A bonded metal–polymer assembly can be considered as a multi-layer structure as shown in the diagram in Figure 6.26. With the exception of gold, every metallic substrate (zone①) is, in normal conditions of use, covered with an oxy-hydroxide film (zone②) resulting from spontaneous reactions with the environment or specific surface treatments. Close to the metallic surface is a complex polymeric zone known as the interphase (zone③), which has a chemical structure and a heterogeneous morphology which differs significantly from those of the adhesive zone (zone④).

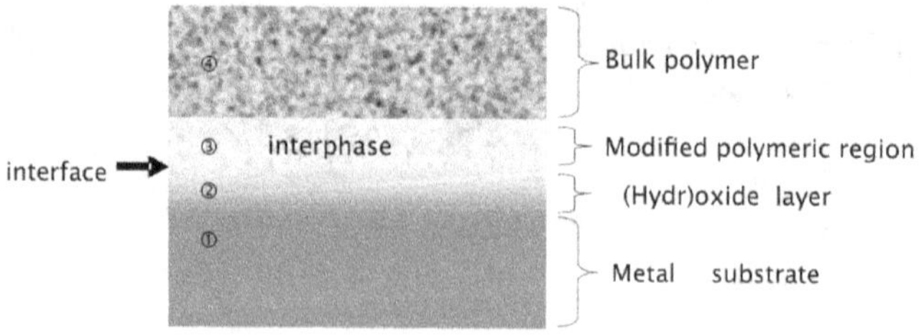

Figure 6.26. *Schematic representation of a polymer/metal joint*

The nature and properties of the interphase play a key role in the final bond performance (practical adhesion, mechanical properties, durability, etc.) and it is therefore important to understand the mechanisms which lead to its formation. In the case of epoxy-amine adhesives, several authors have highlighted an enrichment in the amine hardener in the vicinity of the metal surface, thus suggesting that this hardener could play a specific role in the formation of the interphase.

Since epoxy-amine–aluminum joints are commonly used in the aeronautic industry, Mercier *et al.* [MER 08] studied the interaction mechanisms between a 1050 aluminum sheet with a high aluminum content (>99.5 mass %) and either a molecule with a single amine function (propylamine CH_3-CH_2-CH_2-NH_2, PA) or a molecule with two amine functions (1,2-diaminoethane NH_2-CH_2-CH_2-NH_2, DAE) which can model a diamine hardener.

Section 6.3.2 showed that 1,2-diaminoethane can interact with both Lewis acid sites (metal cations) on the surface of a 1050 aluminum alloy to form O-Al...N complexes and with OHs present at the surface causing the protonation of the DAE. Mercier *et al.* showed that propylamine leads to the same type of interactions [MER 08].

Figure 6.27 shows the evolution of the aluminum concentration in a xylene solution containing 0.5 $mol.L^{-1}$ PA or DAE as a function of the contact time with the 1050 aluminum surface, measured by atomic emission spectroscopy (ICP-AES). The appearance of aluminum in the solution containing the DAE is attributed to partial dissolution of the oxy-hydroxide film covering the aluminum substrate. Although the two molecules (PA and DAE) interact in a similar way with the reactive surface sites, only the DAE is able to dissolve the surface. The dissolving mechanism is therefore linked to the presence of two amine terminations on the molecule, helping to form a bidentate-mononuclear complex (i.e. a complex resulting from the adsorption of both amine terminations of the molecule on the same metal cation), which weakens the bonds of this cation with the oxide matrix and allows it to pass into the solution. It is interesting to highlight the fact that the amine hardeners used in adhesives have at least two amine functions.

Similar dissolution mechanisms linked to the formation of chelate complexes between the surface and the ligands in contact with the surface can also be suggested in order to assess corrosion mechanisms (dissolving of the passive film) or in geochemistry (dissolving of minerals in natural water).

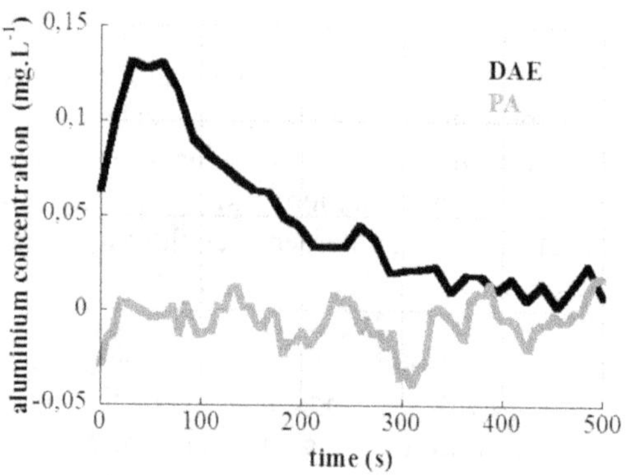

Figure 6.27. *Evolution in aluminum concentration in a xylene solution containing 0.5 mol.L^{-1} PA or DAE as a function of the contact time with the 1050 aluminum surface. Taken from [MER 08]. copyright (2008) with permission from Elsevier*

6.4.3. *Heterogeneous catalysis*

For decades, the acid-base concept has been widely used in the domain of heterogeneous catalysis. For instance, solid acid catalyzers play a key role in refining processes. Although not used as widely, solid base catalyzers are important in the conversion of biomass for example.

Several industrial processes use bi-functional catalyzers, which have both acid and base surface sites. Their interaction with the various types of surface sites can therefore occur simultaneously or successively, with one or more molecules. Figure 6.28 shows a typical example of a simultaneous reaction with a single molecule: dehydration of 1-butanol on an alumina doped with Na. One of the hydrogens of the -CH_2- group will react with the base surface sites (B), while the oxygen will interact with the acid sites (A) via its electron pair.

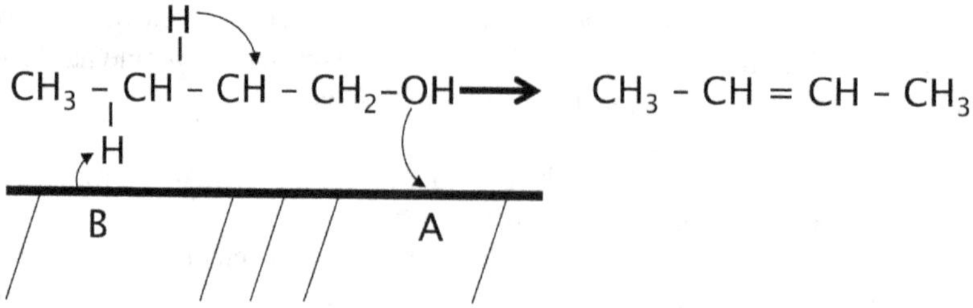

Figure 6.28. *Dehydration mechanism of 1-butanol on Na-doped alumina*

7

Oxide–Solution Interfaces: Surface Charges

We will briefly assess the origin of the charges existing at a hydrated oxide–aqueous solution interface and how the knowledge of the pH of the solution in contact with it, which corresponds to a zero overall charge (PZC), allows us to characterize the acid-base behavior of this type of surface. The PZC represents an important parameter which, for example, governs the adhesion of the surface under investigation with a polymer coating.

When brought into contact with the environmental atmosphere, a material is covered with an oxide layer and several hydration layers, depending on the atmosphere hygrometry. The water molecules in contact with the oxidized substrate dissociate into protons and hydroxyl groups, which establish strong chemical links with the constituents of the oxide according to:

$$M_2O + H_2O \rightarrow M_2O + H^+ + OH^- \rightarrow 2MOH$$

Most of the time, the surface of a metal oxide will present M-OH hydroxyl sites, which are also capable of generating acid-base interactions which will be beneficial, for example, in the case of adhesion with polar polymers.

Therefore, a means of characterizing and predicting the behavior of these sites is using an amphoteric electrolyte to reveal the acid-base properties of such surfaces.

First, the key parameters defining the acid-base character of a surface will be recalled, as well as the mechanisms involved when such surfaces are brought into contact with an electrolyte.

7.1. Brønsted acidity and basicity

Relationship between the surface charge and the absorptions.

A hydroxylated surface can either capture or release protons according to an acidic reaction:

$M-OH \rightarrow M-O^{-}+H^{+}$ with an equilibrium constant K_1

or according to a basic reaction:

$M-OH+H^{+} \rightarrow M-OH_2^{+}$ with an equilibrium constant K_2

When an oxidized surface is brought into contact with an amphoteric electrolyte, the M-OH complexes will be able to dissociate and confer a basic, neutral or acid character to the substrate.

In this way, the amphoteric electrolyte (H_2O most of the time) is used to reveal acid-base potentialities and an overall surface charge which may be positive, negative or zero, depending on the ionization of the hydroxyl sites.

Thus, at the surface scale, an overall superficial charge density can be defined by:

$$\sigma_0 = F\left[\Gamma_{MOH_2^+} - \Gamma_{MO^-}\right] \qquad [7.1]$$

where F refers to the Faraday constant (96,500 $C.mole^{-1}$) and Γ_i corresponds to the Gibbs adsorption of the i constituent on the surface (excess number of moles per unit area).

The overall charge σ_0 depends on the nature of the surface hydroxyls and on the pH of the aqueous solution. It gives information on the extent of dissociation (nature and number of sites involved).

7.2. Point of zero charge (PZC)

Definition and relationship with the equilibrium constant.

The acido-basicity of a given oxide is measured by its point of zero charge (PZC). The PZC corresponds to the pH of the electrolyte for which σ_0 is zero.

The surface charge σ_0 and the associated electrostatic potential ϕ_0 are therefore positive when the pH of the solution is less than the PZC of the oxide and negative when the pH of the solution is larger than the PZC of the oxide.

The PZC of the oxide is linked to the equilibrium constants K_1 and K_2 by a simple calculation.

$$K_1 = \left|H^+\right|\left|MO^-\right| \text{ and } K_2 = \left|MOH_2^+\right| / \left|H^+\right|$$

(the activity of $|MOH|$ is taken as equal to 1).

Defining the PZC as $PZC = -\log |H^+|_{PZC}$ as definition, we write that the charge is zero on the surface either:

$$\left|MO^-\right|_{PZC} = \left|MOH_2^+\right|_{PZC}$$

Or $K_1 / K_2 = \left|H^+\right|^2 \left|MO^-\right| / \left|MOH_2^+\right| = \left|H^+\right|^2_{PZC}$ hence

$$\frac{1}{2}\log K_1 / K_2 = \log\left|H^+\right|_{PZC} \text{ and } \frac{1}{2}\,(pK_1 - pK_2) = PZC \qquad [7.2]$$

(with $pK = -\log K$).

The points of zero charge of a number of oxides and hydroxides were measured by Parks [PAR 65]. Some characteristic values are reported in Table 7.1.

The PZC is a parameter directly linked to the nature of the oxide surface and depends on several factors such as the charge/radius ratio of the metal cation, the hydration degree as well as the crystalline structure of the oxide.

Oxides	PZC
Al_2O_3 (thermal oxidation)	6.7
Al_2O_3 (hydrated)	9.2
Al_2O_3 (amorphous)	7.5, 8.0
Cr_2O_3 (hydrated)	7.9
Fe_2O_3	6.5, 6.9
$FeOOH$	6.5, 6.9
Fe_2O_3 (hydrated)	8.5
TiO_2	4.7
TiO_2 (hydrated)	6.2
SiO_2 (quartz)	2.2
SiO_2 (gel)	1.8
MgO	12

Table 7.1. *Some PZC values of oxides in water. From Parks [PAR 65]*

A more exhaustive analysis of the PZC can be found in the Jolivet book [JOL 94].

Without going into details of a formalism that is highly dependent on writing conventions for the acid-base reactions, one can qualitatively use the characteristic values of the PZC of oxide surfaces and values of the acidity or basicity constants (K_a or K_b) of various polymer coatings to qualify the adhesion of the coating [BOL 68, NOL 65].

For example, when spread onto an acid like SiO_2, the polymers containing an amino basic group ($R\text{-}NH_2$) or a carboxylic group (R-COOH) will lead to a strong adhesion.

Likewise, on a basic surface like MgO, the polymers containing phenol or carboxylic acid will lead to good adhesion.

Good adhesion is conditioned by the difference Δ = PZC-pK_a or, depending on the polymer nature Δ = pK_b-PZC, which must be largely positive thus measuring the strength of the acid-base bonds.

In the expressions of Δ, K_a represents the acid dissociation constant of the polymer and K_b its basic dissociation constant [BOL 68].

When the parameter Δ is negative or zero, the dispersion forces are much stronger than the acid-base interactions.

7.3. The oxide–solution interface

Recalling the description of the oxide–solution interface compared to the metal–solution interface. Electrochemical double layer: Helmholtz and Stern models (Helmholtz layer + Gouy-Chapman diffuse layer).

The surface charge in contact with a solution can be produced in two ways:

– by ionization of a surface group, for example dissociation of a proton of a carboxylic group (-COOH$\rightarrow$ COO^- + H^+) which leaves a negatively charged surface;

– by adsorption of ions in solution which can be adsorbed on a initially neutral or on sites of opposite charges, for example cations Na^+ on anionic sites COO^-.

Whatever the mechanism is, the final surface charge is balanced by a region composed of charged ions of opposites signs, localized in the immediate vicinity of the surface (Helmholtz layer) and in a diffuse layer (Gouy Chapman layer) and subjected to a thermal agitation as indicated in Chapter 6.

Figures 7.1 and 7.2 present the distribution of ions and the evolution of the potential in the case of a metal–solution and an oxide–solution interfaces, respectively.

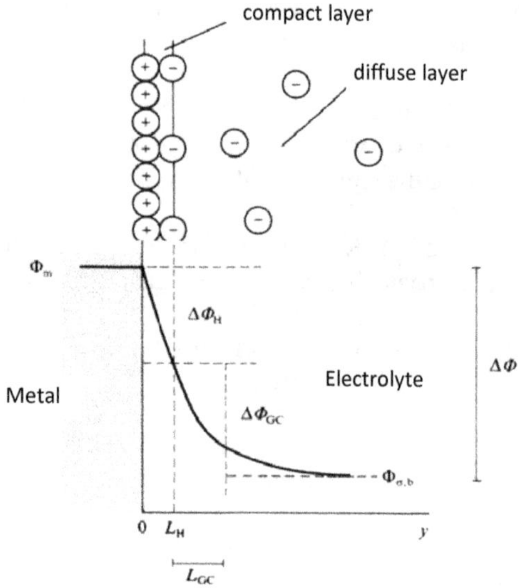

Figure 7.1. *Metal–solution interface. Ion distribution and evolution of the potential. The compact (Helmholtz) double layer is in contact with the surface; the diffuse (Gouy–Chapman) layer lies beyond*

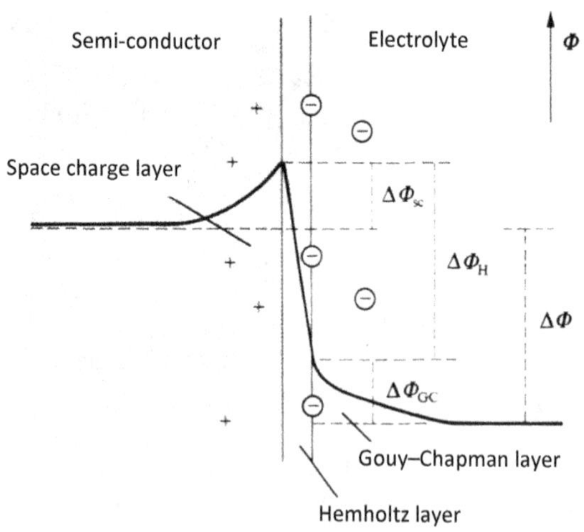

Figure 7.2 *Oxide semiconductor–solution interface. Ion distribution and evolution of the potential. The difference with the previous figure lies in the existence of a space charge layer in the substrate*

The difference between the two figures comes from the existence of a space charge layer in the case of an oxidized surface. In the latter case, the surface charges are spread over a certain thickness linked to the semiconductor character of the oxide. At finite temperature, the forbidden band which separates the valence and conduction bands is no more impermeable. Some electrons can pass into the conduction band thus leaving some positive holes in the valence band, contributing to the overall electric conduction.

In both cases, the overall electroneutrality is written as:

$$\sigma_0 + \sigma_d = 0$$

where σ_0 is the surface charge (which integrates a possible contribution of the oxide layer) and σ_d the charge in the solution.

The charge distribution in the diffused layer corresponds to a Boltzmann distribution:

$$\rho_x = F \Sigma_i Z_i (C_i)_b \exp(-Z_i F \Phi_x / RT)$$

Φ_x represents the potential in the layer at position x relatively to that which is found in the core of the solution ($x \rightarrow \infty$), which will be taken by convention=0, F the Faraday is equal to 96,500 C, Z_i the charge of ion i, $(C_i)_b$ the volume concentration of ion i, T the temperature and R the perfect gas constant.

For a symmetrical binary electrolyte of concentration C_b [LAN 93]:

$$\rho_x = ZFC_b \left[\exp(-ZF\Phi_x / RT) - \exp(ZF\Phi_x / RT)\right] \quad [7.3]$$

where Z is the positive charge of the cation in solution and -Z is that of the anion, Φ_x the potential in the diffused layer always relatively to the core of the solution taken by convention=0.

The Poisson equation links the charge density to the potential (be careful with the written convention of this law when used by different authors):

$$\rho_x = -\varepsilon\varepsilon_0 \frac{d^2\Phi}{dx^2} \quad [7.4]$$

where ε_0 is the vacuum permittivity (8.854 $10^{-12}\,C^2 J^{-1} m^{-1}$) and ε is the relative permittivity.

The integration of this equation, together with the use of the Gauss theorem, leads to the expression of the charge in the diffuse layer and of its capacity. The details of this classic calculation can be found for example in Volume 2 of the book by Bockris and Reddy [BOC 70].

$$\sigma_d = -\varepsilon\varepsilon_0 \left(\frac{8RTC_b}{\varepsilon\varepsilon_0} \right)^{1/2} \sinh\left(ZF\Phi_0 / 2RT \right) \qquad [7.5]$$

$$C_{GC} = \frac{dq}{d\phi} = \varepsilon\varepsilon_0 \left[\frac{2Z^2F^2C_b}{\varepsilon\varepsilon_0 RT} \right]^{1/2} \cosh\left(\frac{ZF\Phi}{2RT} \right) \qquad [7.6]$$

Using the same type of calculation, the variation in the potential Φ as a function of the distance x to the interface can be calculated. Approximating $\sinh x \approx x$ (in the case of a potential $\Phi_0 > 0$ and a surface charge $\sigma_0 > 0$):

$$\Phi(x) = \Phi_0 \exp(-x / L_{GC}) \qquad [7.7]$$

where L_{GC} is the thickness of the Gouy Chapman layer given by the relationship:

$$L_{GC} = \left[\frac{\varepsilon\varepsilon_0 RT}{2Z^2F^2C_b} \right]^{1/2} \qquad [7.8]$$

The variation in the potential hence of the charges in the diffuse layer is exponentially decreasing.

The Debye Huckel theory applied to a charge cloud with a total value of σ_d shows that the electric effect of this charge cloud on a reference ion can be simulated by placing the whole charge σ_d on a plane located at the Debye distance $\kappa^{-1} = L_{GC}$ of the reference ion (Bockris and Reddy [BOC 70]).

Equation [7.8] shows that the Debye length $\kappa^{-1} = L_{GC}$ increases for weakly concentrated solutions. For a 10^{-4} mole/L solution containing ions with charge Z=1, at a temperature of 298 K and having a relative dielectric constant $\varepsilon = 80$ we get:

$$\kappa^{-1} = L_{GC} = 30 \text{ nm}.$$

This value is much higher than the distance L_H characterizing the Helmholtz layer which is a few tenths of nanometers.

In the case of the metal/electrolyte interface (Figure 7.1), the capacity of the system corresponds to two capacities connected in series (Helmholtz layer and Gouy layer). It is the Stern model.

$$\frac{1}{C} = \frac{1}{C_H} + \frac{1}{C_{GC}}$$

For a concentrated electrolyte (0.5–1 mole/L), C_{CG} increases and the previous relationship is reduced to:

$$\frac{1}{C} = \frac{1}{C_H}$$

i.e. $C \simeq C_H$

The majority of the charges in the solution are localized on the external Helmholtz plane and very few charges are in the diffused zone. The Helmholtz model of two charged planes can be applied (simple condenser):

$$\sigma_0 = -\sigma_H \text{ with } \Phi_0 - \Phi_{Sol} = \frac{L_H}{\varepsilon\varepsilon_0}\sigma_0 \qquad [7.9]$$

where L_H is the distance between the surface and the Helmholtz external plane.

For a weakly concentrated electrolyte the situation is reversed:

$$\frac{1}{C} = \frac{1}{C_{GC}}$$

i.e. $C \simeq C_{CG}$.

The charges in the solution are diffused under the action of the electric field and thermal agitation.

In the case of the oxide (semiconductor)/electrolyte interface (Figure 7.2) the capacity of the double layer corresponds to three capacities connected in series:

$$\frac{1}{C}=\frac{1}{C_{sc}}+\frac{1}{C_H}+\frac{1}{C_{GC}}$$

In general the term, $1/C_{sc}$ is dominant (weak number of charge carriers) and [LAN 93] : $\frac{1}{C}=\frac{1}{C_{sc}}$

In this way, the measurement of the capacity of the semiconductor oxide/electrolyte system provides information on the space charge layer in the oxide.

Similarly to the calculation carried out for the diffuse layer in the electrolyte, whose details are also given in Volume 2 of the book by Bockris and Reddy [BOC 70], the following equations can be written:

$$q_{sc}=\frac{2\varepsilon\varepsilon_0 RT}{FL_{sc}}\sinh\left(\frac{F\Delta\Phi_{sc}}{2RT}\right) \quad [7.10]$$

$$C_{sc}=\frac{dq_{sc}}{d\Phi_{sc}}=\frac{\varepsilon\varepsilon_0}{L_{sc}}\cosh\left(\frac{F\Delta\Phi_{sc}}{2RT}\right) \quad [7.11]$$

With $$L_{sc}=\left(\frac{\varepsilon\varepsilon_0 RT}{2F^2 C_{sc}}\right)^{1/2} \quad [7.12]$$

$\Delta\Phi_{sc}$ represents the difference in potential between the inner part of the semiconductor and the surface. The Debye length L_{sc} involves the concentration in intrinsic carriers of the semiconductor ($C_{sc}=C_n=C_p$).

The calculation is more complicated in the case of semiconductors containing dopants [PLE 80, MOR 80].

The extrinsic semiconductors present a gap between the number of carriers with positive and negative charges. In a n-type semiconductor, the concentration of the electrons greatly exceed that of the holes ($C_n >> C_p$). For a p-type semiconductor, the situation is opposite ($C_p >> C_n$).

This type of situation is obtained in a classic semiconductor like silicon by adding electron acceptor dopants like boron or electron donors like phosphorous.

An excess or lack of oxygen ions can lead to a similar effect in oxides. Oxides such as ZnO, TiO_2, V_2O_5, are n-type whereas Cr_2O_3, NiO, Cu_2O are p-type [LAN 93].

When the concentration in the majority carriers in the space charge layer is less than in the bulk material (inversion layer), the Mott–Schotky equation is obtained .

The variation of the capacity with the potential can be written in the case of a n-type semiconductor:

$$\frac{1}{C_{sc}^2} = \left(\frac{2L_{sc}}{\varepsilon\varepsilon_0}\right)^2 \left(\frac{F\Delta\Phi_{sc}}{RT} - 1\right) \qquad [7.13]$$

With $L_{sc} = \left(\dfrac{\varepsilon\varepsilon_0 RT}{2F^2 C_n}\right)^{1/2}$

The analysis of the variation of the capacity with the imposed potential allows us to determine the flat band potential, i.e. the potential of the electrode corresponding to $\Phi_{Surf} = \Phi_{bulk}$, and the charge density in the region of space charges.

7.4. Electrocapillarity in the oxide–solution interface

Evolution of the solid–liquid interfacial tension analyzed from the Gibbs adsorption equation. Determination of the PZC and of the interface maximum charge.

7.4.1. *Evolution of the interfacial tension*

By placing a drop of acidic or basic aqueous solution on the oxide surface under investigation, the solid–liquid interfacial tension is modified due to adsorption or desorption of ions found in the solution, following the mechanisms already seen.

The influence on the interfacial tension of the new interfacial composition on the interfacial tension, at a constant temperature and near the equilibrium, can be written with the help of the Gibbs adsorption relationship (equation [1.10]) :

$$d\gamma_{SL} = -\Sigma_i \Gamma_i d\bar{\mu}_i \qquad [7.14]$$

with Γ_i adsorption of constituent i in the interface and $\bar{\mu}_i$ the electrochemical potential of constituent i.

The electrochemical potential $\bar{\mu}_i$ corrects the classic chemical potential μ_i with a term corresponding to the electrostatic energy of an ion with the charge Z_i in the local potential Φ :

$$d\bar{\mu}_i = RTd\ln C_i + Z_i F d\Phi \qquad [7.15]$$

Let us consider a simple situation where a BA solid oxide (of M-OH type) is dissociated in the presence of an electrolyte containing an AY acidic species (of HCl type).

Let us ignore the possibility of specific adsorption on the surface of other ions originating from the solution.

In this classic simple case, the common ion which governs the equilibrium in the interface is A=H.

The dissociation equilibrium of the oxide is written as:

$$AB \rightleftarrows A^+ + B^- \qquad [7.16]$$

while the acid in solution is completely dissociated. $AY \rightarrow A^+ + Y^-$

The [7.14] relationship is written as:

$$d\gamma_{SL} = -\Gamma_{A+} d\bar{\mu}_{A+} - \Gamma_{B-} d\bar{\mu}_{B-} - \Gamma_{Y-} d\bar{\mu}_{Y-}$$

As $d\bar{\mu}_{AB} = d\bar{\mu}_{A+} + d\bar{\mu}_{B-} = 0$ (equilibrium [7.16])

$$d\gamma_{SL} = -(\Gamma_{A+} - \Gamma_{B-}) d\bar{\mu}_{A+} - \Gamma_{Y-} d\bar{\mu}_{Y-} \qquad [7.17]$$

The electroneutrality condition of the interface charges is written, in a simple Helmholtz layer model:

$$\sigma_0 / F = (\Gamma_{A+} - \Gamma_{B-}) = (0 - \Gamma_{Y-})$$

i.e.:

$$d\gamma_{SL} = -\frac{\sigma_0}{F} d\bar{\mu}_{A+} - \frac{\sigma_0}{F} d\bar{\mu}_{Y-} \qquad [7.18]$$

The Y^- ions are adsorbed onto the Helmholtz external plane where they are subjected to the electrostatic potential Φ_d. At equilibrium, we have the equality of the electrochemical potential of the Y^- species on the surface and in the core solution where $\left|Y^-\right| = \left|A^+\right|$ and a constant potential Φ_L is observed.

Hence

$$(d\bar{\mu}_{Y-})_{ads} = (d\bar{\mu}_{Y-})_{sol} = RTd\ln\left|A^{+}\right|_{sol} + Fd\Phi_L = RTd\ln\left|A^+\right|_{sol} \qquad [7.19]$$

The adsorption equilibrium of A^+ species on the surface can be written, in the framework of a Langmuir model:

$$A^+_{sol} + V_{surf} \rightleftarrows A^+_{surf} \text{ with an equilibrium constant } K = \frac{\theta}{(1-\theta)} \frac{1}{\left|A^+\right|_{sol}}$$

θ represents the fraction of surface sites occupied by A^+ and V_{surf} the fraction of empty sites.

The electrochemical potential of A^+ species on the surface subjected to the local potential Φ_0 is therefore written [BEN 83]:

$$d\bar{\mu}_{A+} = RTd\ln\frac{\theta}{(1-\theta)} + Fd\Phi_0 = RTd\ln\left|A^+\right|_{sol} + Fd\Phi_0 \qquad [7.20]$$

Injecting equations [7.19] and [7.20] in expression [7.18] of the solid–liquid interfacial tension variation leads to the following relationship:

$$d\gamma_{SL} = -2\frac{\sigma_0}{F}RTd\ln\frac{\theta}{(1-\theta)} - \sigma_0 d\Phi_0 = -2\frac{\sigma_0}{F}RTd\ln\left|A^+\right|_{sol} - \sigma_0 d\Phi_0 \quad [7.21]$$

On the one hand, this relationship shows that the solid–liquid tension is maximum for the PZC where σ_0 and Φ_0 are zeros, and on the other hand that there is a chemical and an electrostatic contribution to the decrease of the interfacial tension.

To integrate equation [7.21], we will use expression [7.5] of $\sigma_0 = -\sigma_d$ (Grahame equation) and the following expression of θ, the normalized surface fraction as shown by Stol [STO 80]:

$$\theta = \frac{\sigma_{max} + \sigma}{2\sigma_{max}} \text{ with } \theta = 1 \text{ for } \sigma = \sigma_{max} \text{ and } \theta = 1/2 \text{ for } \sigma = 0$$

with $C_b = \left|A^+\right|_{sol}$

The integration is between the PZC where the surface charge and potential are zero and the current point characterized by the charge σ_0 and the potential Φ_0

$$\gamma_{SL} - \gamma_{SL}^0 = -\int_0^{\Phi_0} (8\varepsilon\varepsilon_0 RTC_b)^{1/2} \sinh(F\Phi/2RT)d\Phi - \int_0^{\sigma_0} 2\sigma\frac{RT}{F}d\ln\frac{\sigma_{max}+\sigma}{\sigma_{max}-\sigma}$$

hence

$$\Delta\gamma_{SL} = -\frac{8RTC_b}{\kappa}\left[\cosh(\frac{F\Phi_0}{2RT}) - 1\right] + \frac{2RT}{F}\sigma_{max}\ln(1 - \frac{\sigma_0^2}{\sigma_{max}^2}) \qquad [7.22]$$

with $\kappa^{-1} = \left[\dfrac{\varepsilon\varepsilon_0 RT}{2F^2 C_b}\right]^{1/2}$ the Debye length.

1) Equation [7.22] may be simplified using a limited development in the vicinity of the PZC where σ_0 and Φ_0 are zero. Moreover, we will leave the second term of equation [7.22] by assuming $\sigma_0 < \sigma_{max}$ and therefore $\sigma_0^2 / \sigma_{max}^2 \approx 0$.

The decrease in γ_{SL} is essentially due to the electrostatic contribution.

Moreover, the potential Φ_0 can be deduced from the difference of the electrochemical potential of the proton applied at the current point and at the point of zero charge knowing that Φ_0 is in fact the difference between the surface potential and the potential in the core of the solution ([JOL 94] and remark 1).

$$\Phi_0 \approx \frac{RT}{F} \ln \frac{\left|H^+\right|}{\left|H^+\right|_{pzc}} \approx 2.3 \frac{RT}{F}(PZC - pH)$$

Hence in the vicinity of the PZC:

$$\Delta\gamma_{SL} = \gamma_{SL} - (\gamma_{SL})_{PZC} \approx -5.3 \frac{RTC_b}{\kappa}(PZC - pH)^2 \qquad [7.23]$$

γ_{SL} varies in a parabolic way around the maximum value obtained at the PZC.

2) Far from the PZC, the surface charge becomes saturated ($\sigma_0 = \sigma_{max}$). The surface potential Φ_0 can no longer vary. Relationship [7.21] can be simplified (the common A^+ ion is the proton):

$$d\gamma_{SL} == -2\frac{\sigma_{max}}{F} RTd\ln\left|A^+\right|_{sol} - \sigma_{max} d\Phi_0 = -2\frac{\sigma_{max}}{F} RTd\ln\left|H^+\right|_{sol}$$

$$\text{hence } \frac{d\gamma_{SL}}{dpH} = 2\sigma_{max}\frac{2.3RT}{F} \quad [7.24]$$

The analysis discussed above leads to two results: the parabolic variation of the interfacial tension in the vicinity of the PZC and the determination of the maximum surface charge far from the PZC as indicated in Figure 7.3 [JOU 98].

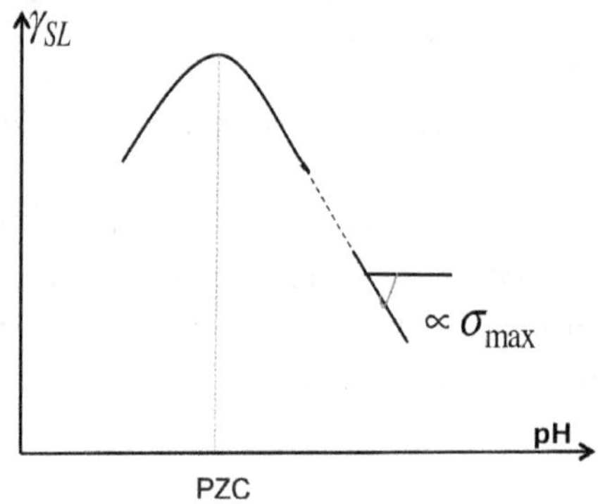

Figure 7.3. *Diagram representation of the solid–liquid interfacial tension variation with the pH of the solution*

REMARK 7.1.– When the proton is the common ion, it follows a Boltzmann distribution between the surface and the bulk solution:

$$\left|H^+\right|_{surf} = \left|H^+\right|_{sol} \exp\left(-F\Phi_0 / RT\right)$$

with Φ_0 the surface potential relatively to the potential of the solution taken equal to 0. Hence:

$$\Phi_0 = \frac{RT}{F}\ln\frac{\left|H^+\right|_{sol}}{\left|H^+\right|_{surf}} \quad [7.25]$$

Relationship [7.25] applied at the PZC is written as:

$$\Phi_0 = 0 \text{ and } \left|H^+\right|_{surf}^{PZC} = \left|H^+\right|_{sol}^{PZC} \qquad [7.26]$$

Equation [7.25] can be rewritten as:

$$\Phi_0 = \frac{RT}{F}\ln\frac{\left|H^+\right|_{sol}}{\left|H^+\right|_{sol}^{PZC}} + \frac{RT}{F}\ln\frac{\left|H^+\right|_{surf}^{PZC}}{\left|H^+\right|_{surf}}$$

or, by introducing the normalized surface fraction: $\theta = \frac{\sigma_{max} + \sigma}{2\sigma_{max}} = 1/2 + 1/2\frac{\sigma}{\sigma_{max}}$

$$\Phi_0 = 2.3\frac{RT}{F}(PZC - pH) + \frac{RT}{F}\ln\frac{1/2}{\theta} \qquad [7.27]$$

The second-term of equation [7.27] is written as:

$$\frac{RT}{F}\ln\frac{1/2}{\theta} = \frac{RT}{F}\ln\frac{1/2}{1/2+\delta\theta} = -\frac{RT}{F}\ln(1+2\delta\theta) \approx -\frac{2RT}{F}\delta\theta \approx -\frac{RT}{F}\frac{\sigma}{\sigma_{max}}$$

and can be neglected when $\sigma << \sigma_{max}$.

The first result (equation [7.23]) can also be deduced directly from the Gibbs adsorption equation and the Helmholtz model for the double layer (single condenser).

$$d\gamma_{SL} = -\Sigma_i \Gamma_i d\bar{\mu}_i$$

hence in the vicinity of the PZC, at constant composition:

$$(d\gamma_{SL})_{PZC}^{c.const} = -\sigma d\Phi \text{ (Lippmann equation)}$$

The simple condenser model links the surface charge to the potential difference:

$$\Phi\text{-}\Phi_{sol} = \Phi\text{-}\Phi_{PZC}$$

through the capacity per unit area $C = \frac{\varepsilon\varepsilon_0}{d}$.

Hence:

$$(d\gamma_{SL})^{c.const}_{PZC} = -\frac{\varepsilon\varepsilon_0}{d}(\Phi-\Phi_{PZC})d\Phi$$

Hence

$$(\gamma_{SL}) = \gamma_{SL}^{PZC} - \frac{\varepsilon\varepsilon_0}{2d}(\Phi-\Phi_{PZC})^2 \qquad [7.28]$$

By setting that the distance d in the definition of the capacity of the single condenser is given by the Debye length:

$\kappa^{-1} = \left[\frac{\varepsilon\varepsilon_0 RT}{2F^2 C_b}\right]^{1/2}$, equation [7.28] is similar to equation [7.23].

Both analyses therefore lead to the same parabolic variation of γ_{SL} in the vicinity of the PZC. Moreover, the first analysis allows us to make an evaluation of the maximum charge of the surface far from the PZC and to easily determine the charge sign on both sides of the PZC $\sigma < 0$ for pH $> PZC$ and $\sigma > 0$ for pH $< PZC$).

Examples of application are given in Chapter 8.

8

Electrocapillarity Applications

The first part that will be considered and illustrated involves the determination of the characteristics of an oxide-solution interface, namely the PZC and the maximum charge.

The second part involves the classic influence of surface charges on the free energy of a prewetting film in contact with an ideal surface. This influence can be modified in the case of a hydrophilic rough surface.

The existence of a prewetting film on a rough oxide surface presenting a charge at the oxide-solution interface can be stabilized, which can lead to persistence of superhydrophilic properties on some oxides.

Finally, the third part deals with the modification of the contact angle by an imposed potential (electrowetting) and evokes some applications in a rapidly developing area.

8.1. Study based on the pH of the oxide surfaces

The evolution of the measured contact angle as a function of the pH of the aqueous solution facilitates the determination of the PZC and of the surface maximum charge.

In the case of a simple experience of wettability, Young's relationship is written as:

$$\gamma_{SV} = \gamma_{SL} + \gamma_{LV} \cos\theta$$

When the pH of the aqueous solution varies, we can state:

$$\frac{d\gamma_{SV}}{dpH} = 0$$

and taking into account the non-segregation of the protons at the liquid/vapor interface [DEF 66]

$$\frac{d\gamma_{LV}}{dpH} \approx 0$$

Hence,

$$\frac{d\gamma_{SL}}{dpH} = \gamma_{LV} \sin\theta \frac{d\theta}{dpH} \qquad [8.1]$$

The PZC of the surface is therefore obtained for $\frac{d\theta}{dpH} = 0$.

The maximum charge, far from the PZC, can be written as:

$$\frac{\sigma_{\max}}{F} = \frac{\gamma_{LV} \sin\theta}{4.6RT} \frac{d\theta}{dpH} \qquad [8.2]$$

In the case of experiments conducted by the two liquid method, equations [8.1] and [8.2] are preserved by simply replacing the liquid vapor tension by the liquid alkane tension.

The first example presented below comes from Vittoz [JOU 98]. First of all, the study involves two model surfaces namely SiO_2 (thermal oxidation of a silicon single crystal) and Al_2O_3 (sapphire).

The contact angle measurements are conducted by the two liquid method in the presence of octane.

For each aqueous solution used, the pH value is checked before and after each angle measurement and the angle value selected is an average out of 10 drops deposited on the substrate. The surfaces studied were cleaned prior to measurements (acetone and drying in dry air) and their cleanliness verified by XPS analysis. The results are presented in the following figures.

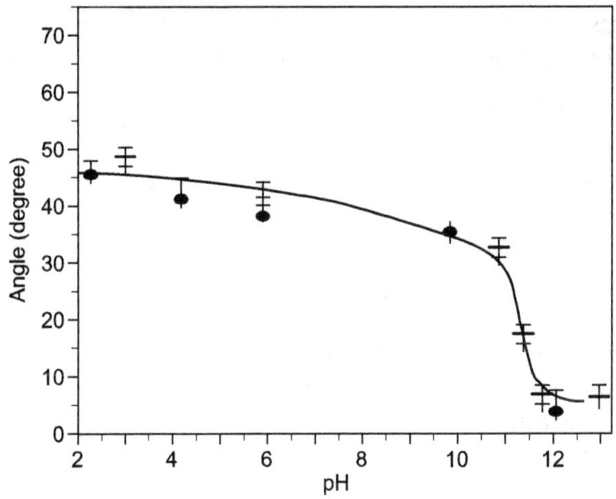

Figure 8.1. *Variation of the contact angle with the pH for a SiO_2 surface. Two liquid method with octane*

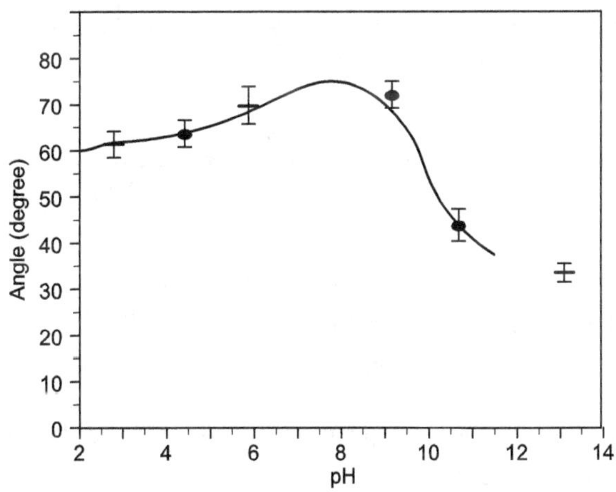

Figure 8.2. *Variation of the contact angle with the pH for a Al_2O_3 surface. Two liquid method with octane*

The sapphire surface shows an amphoteric oxide behavior with a value of PZC close to 8. On either side of this PZC value, the curve $\theta = f(pH)$ presents a positive then negative slope corresponding to positive then negative charges.

In the case of the silica surface, we observe a PZC value close to 2 and a negative slope $d\theta / dpH$ at all pH values, which corresponds to negative surface charges (acid behavior). The maximum charge can be calculated (equation [8.2]) at the inflexion point close to pH=11. In the example presented here, the calculated value is $\sigma_{\max} = -8.4\ \mu C / cm^2$.

Atomic force microscopy (AFM) experiments in liquid environment have been carried out on the same type of silica samples.

The force versus distance plots are measured in a liquid cell using a Park Scientific instruments microscope. A Si_3N_4 sphere 50 nm in radius is fixed at the end of the cantilever. The substrate under consideration is the silica sample [VIT 00].

Three pH are considered for the liquid solution (3; 5.7 and 11) with KCl additions to maintain constant (10^{-3} mole/L) the ionic force.

For large distances ($D \geq 50\ nm$) the electrostatic interactions are greater than the attractive Van der Waals interactions (see the following section). They can be represented by:

$$F_{el} = \frac{2\pi\kappa_D^{-1} R}{\varepsilon\varepsilon_0}\left\{(\sigma_t^2 + \sigma_s^2)\exp\frac{-2D}{\kappa_D^{-1}} + 2\sigma_t\sigma_s \exp\frac{-D}{\kappa_D^{-1}}\right\} \qquad [8.3]$$

with ε and ε_0 the relative and absolute permittivities, κ_D^{-1} the Debye length, σ_t and σ_s the surface charges of the sphere Si_3N_4 and of the surface SiO_2 and R the radius of the sphere Si_3N_4.

The plots $F_{el} = f(\text{D})$ corresponding to the three pH of the liquid solution are shown in Figure 8.3.

For the solution at pH 3 – a value that is close to the PZC of silica ($PZC \approx 2.5$) – a surface charge of SiO_2 close to 0 is expected and therefore a negligible F_{el}.

Likewise for the solution at pH 5.7 – a value that is close to the PZC of Si_3N_4 (PZC=6) – a surface charge of Si_3N_4 close to 0 is expected and therefore a negligible F_{el}.

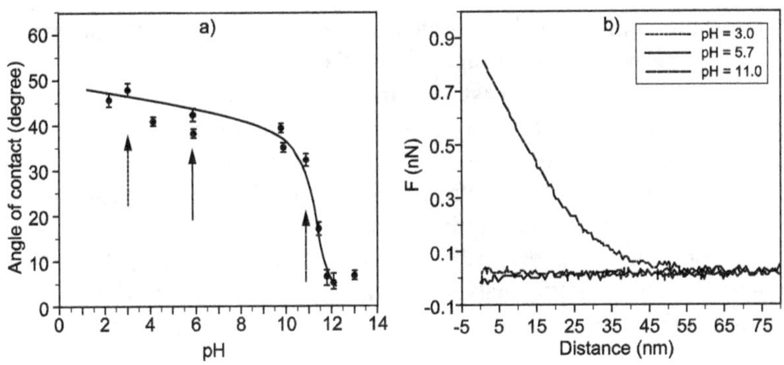

Figure 8.3. *a) Evolution of the contact angle as a function of the pH measured on a silica surface by the 2 liquid method (octane) and b) f(D) curves measured by AFM in solutions with three different pH [DUB 00]. For a color version of the figure, see www.iste.co.uk/joud/surfaces.zip*

At pH=11, a net electrostatic interaction is observed which corresponds to similar charges on the two surfaces. By fitting the experimental plot using equation [8.3], the following values are obtained :

$$\kappa_D^{-1} = 23\ nm\ \sigma_t = -0.4\ \mu C / cm^2\ \sigma_s = -2\ \mu C / cm^2$$

The silica surface charge, obtained on a surface presenting a diameter of 50–70 nm ($-2\ \mu C / cm^2$) must be compared to the value obtained by the contact angles method ($-8.4 \mu C / cm^2$), i.e. an average value obtained on a surface presenting a diameter of 1–2 mm. A fair agreement on the sign and the order of magnitude is observed.

The next example refers to a model alloy Fe-Cr (18.8at%)-Si (2.3 at%) that can simulate the behavior of the passive film formed on the surface of a stainless steel. For this purpose, the alloy samples undergo thermal processing at 820°C for 15 min in a H_2/H_2O atmosphere. Two atmospheres are selected corresponding to two dew points, namely (-56°C) the most oxidizing and (-58°C) the least oxidizing (obtained by changing the water content in atmosphere).

The surface of the samples is then cleaned by plasma Ar processing (50 W–10 min) leading to a weak carbon contamination evaluated by XPS to 2–3 carbonaceous monolayers.

For the two processing atmospheres, the passive film is characterized by XPS. The Fe which mostly appears in the metal form is supposed to be localized mainly under the passive film. The composition of the passive film appears strongly dependent on the annealing atmosphere. For the annealed sample in the least oxidizing atmosphere (-58°C), silicon oxide appears to be the main constituent of the surface (Si^{4+} = 33at%), whereas for the most oxidizing atmosphere (-56°C), the quantities of oxidized iron and especially of chromium oxide appear to be significant (Cr^{3+} =21 at%). In particular, angular dependent XPS analysis confirms an enrichment of the extreme surface in chromium oxide.

For samples corresponding to the two processes, the wettability results obtained by the 2 liquid method (heptane) are presented in Figure 8.4.

A significant evolution of the curves $\theta = f(pH)$ depending on the annealing atmosphere can be observed. The evolution observed on the sample (-58°C) is fully comparable to the curve observed on the thermally oxidized silica. However, the curve obtained on the most oxidized sample (-56°C) shows two bumps, the first bump centered around pH=2.5 corresponding to the PZC of the silica and the second bump centered around pH=7 corresponding to the PZC of the chromium oxide.

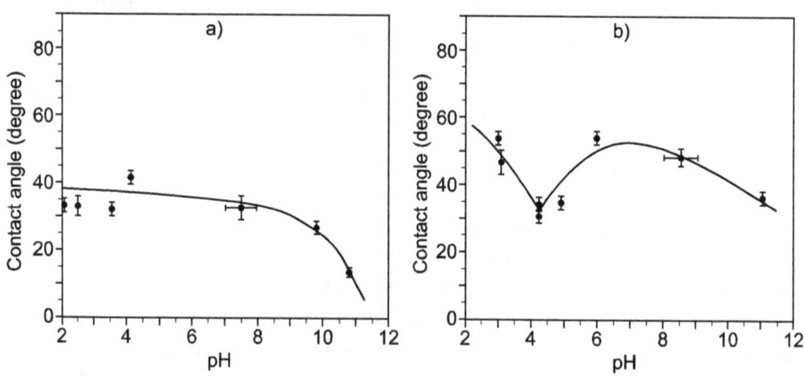

Figure 8.4. *Variation of the contact angle with the pH for an model alloy: Fe-Cr (18.8at%)-Si (2.3at%) having undergone two types of oxidizing processes at 820°C. a) Dew point -58°C (the least oxidizing); b) dew point -56°C (the most oxidizing). From Vittoz [VIT 96]*

8.2. Study of the stability of a liquid film on an oxide surface

8.2.1. *Case of a flat surface*

First of all, let us consider the case of a liquid film (aqueous solution) several tens of nm thick on a flat oxide surface. The liquid–oxide interface presents a double electrochemical layer whose thickness is given by the Debye length κ_D^{-1} (Figure 8.5).

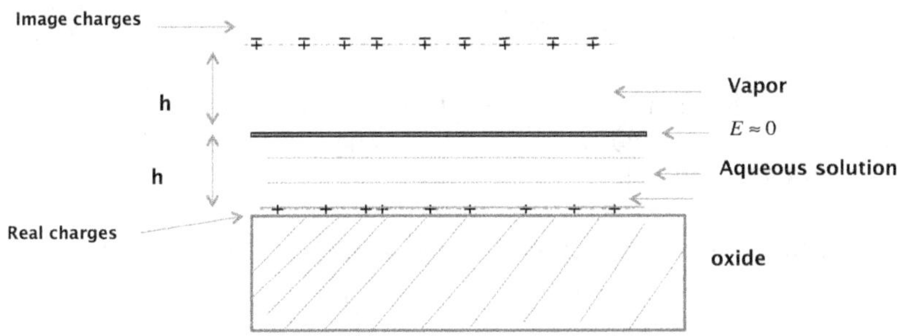

Figure 8.5. *Schematic representation of a liquid film on a flat oxide with real charges and image charges*

The liquid–vapor interface is characterized by an electric field close to zero due to the large difference between the dielectric constants ($\varepsilon_{gas} = 1$ and $\varepsilon_{water} = 80$). In this way, the double electrochemical layer is pushed away by an electrostatic image. This electrostatic repulsion was evaluated for the first time by Langmuir [LAN 38].

The disjoining pressure associated with this electrostatic repulsion gives rise to an electrostatic energy per unit area $G_{el.st}$ corresponding to the interactions between the double layer and its electrostatic image. Supposing that the surface potential of the solid is weak, the electrostatic energy $G_{el.st}$ per unit area can be written for monovalent ions in solution like Na^+ and Cl^- [ISR 11] as:

$$G_{el.st} = \frac{2\sigma^2}{\varepsilon\varepsilon_0}\frac{1}{\kappa_D}\exp - \kappa_D 2h \qquad [8.4]$$

where σ is the surface charge density, ε and ε_0 are the liquid relative dielectric constant and the vacuum dielectric constant, 2h is the distance between the real double layer and its image.

The Van der Waals long distance interactions must also be taken into account in the evaluation of the specific free energy of the liquid film. This latter contribution will be shown to be negligible in view of the thickness considered for the liquid film. The free energy of the film is written as:

$$F = \gamma_{SL} + \gamma_{LV} + G_{el.st} + G_{VW} \qquad [8.5]$$

The two last terms of equation [8.5] correspond to the contributions described in the DLVO theory, which is fully described in Israelachvili's book [ISR 11, VER 48].

The following writing for the Van der Waals contribution is proposed:

$$G_{VW} = -\frac{A_{132}}{12\pi h^2} \qquad [8.6]$$

with A_{132} the Hamaker constant for bodies 1 and 2 interacting through body 3; h the distance between the two interfaces.

The combination relationships lead to:

$$A_{132} \approx (A_{11}^{1/2} - A_{33}^{1/2})(A_{22}^{1/2} - A_{33}^{1/2})$$

For the oxide/water/air system under consideration, the following values can be taken from Israelachvili's book [ISR 11]:

$$A_{11} = A_{oxide} \approx 25.10^{-20} J, A_{33} = A_{water} \approx 3.7.10^{-20} J, A_{22} = A_{air} \approx 0$$

which leads to: $A_{132} \approx -6.10^{-20} J$

The Van der Waals contribution is positive, but for the typical values of h considered (between 10 nm and 70 nm), the value of G_{VW} is negligible ($G_{VW} \approx 10^{-3} G_{el.st}$).

The variation with h of the specific free energy of the film is written when considering equations [8.5] and [8.6]:

$$dF = dG_{el.st} = \frac{-4\sigma^2}{\varepsilon\varepsilon_0} \exp(-2\kappa_D h)dh \qquad [8.7]$$

The film presents a spontaneous tendency to thickening. In fact:

$dF \leq 0$ for $dh \geq 0$.

8.2.2. *Case of a rough surface*

8.2.2.1. *Stability of the prewetting film*

Let us now consider the case of a rough oxide surface. For the sake of simplification, the roughness will be represented by regular slots of the same depth.

Various fillings of the pores by liquid water are discussed in Figure 8.6.

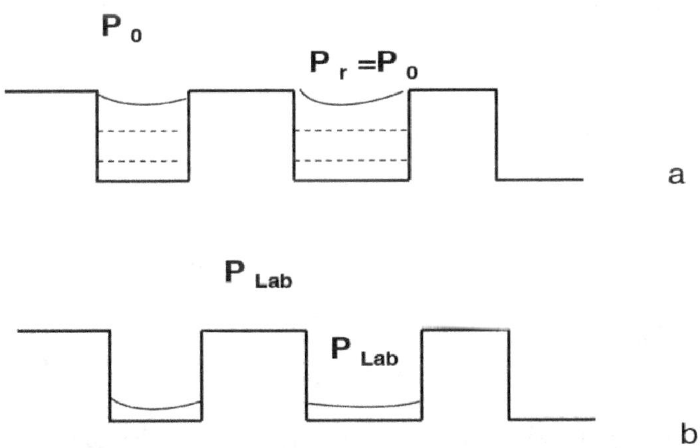

Figure 8.6. *Comparison between two situations: a) the pores are completely filled under an equilibrium pressure P_0; b) the pores are partially filled under an imposed pressure P_{Lab}.*

A consequence of equation [8.7] and of the existence of charges at the liquid–oxide interface is an easy filling of the surface pores which present an equilibrium pressure P_r given by the Kelvin equation [ADA 90] in a standard situation for which the liquid wets the lateral walls of the pores:

$$\frac{2\gamma_{LV}}{r_m} = -\frac{RT}{v_L}\ln\frac{P_r}{P_0}$$

where r_m is the average radius of the pores, v_L is the molar volume of the liquid, P_r and P_0 the partial pressure inside the pore and the partial pressure of water vapor close to a macroscopic drop of solution respectively.

As P_r is less than P_0 in the first situation and taking into account equation [8.7], a complete filling of the surface pores with $P_r = P_0$ in the final situation can be reasonably envisaged.

A second consequence is the capacity of the liquid film to resist vaporization in a dry atmosphere.

As indicated in Figure 8.6, the chemical potential of a vapor mole can be expressed in the two situations:

– in situation a: $\mu_{vap} = \mu_0 + RT\ln\frac{P_r}{P_0} = \mu_0$ when the pore is completely filled;

– in situation b: $\mu_{vap} = \mu_0 + RT\ln\frac{P_{Lab}}{P_0}$ when the external pressure is fixed at P_{Lab}.

Between the two situations we have a variation of chemical potential:

$$\Delta_a^b\mu = RT\ln\frac{P_{Lab}}{P_0}$$

For a negative variation dh of the film thickness, the number of water moles lost by the film per unit area is written as:

$$\eta = \frac{1}{v_L}dh \text{ thus } \eta\Delta_a^b\mu = \frac{RT}{v_L}\ln\frac{P_{Lab}}{P_0}.dh$$

When taking into account a limited vaporization of the film, the variation in the specific free energy of the film is written as:

$$dF = \left\{ \frac{-4\sigma^2}{\varepsilon\varepsilon_0} \exp(-2\kappa_D h) - \frac{RT}{v_L} \ln \frac{P_{Lab}}{P_0} \right\} .dh \qquad [8.8]$$

or $$dF = -\left\{ 2\kappa_D .G_{el.st} + \frac{RT}{v_L} \ln \frac{P_{Lab}}{P_0} \right\} .dh \qquad [8.9]$$

In this simple calculation the variation between the liquid and vapor standard states has been neglected, assuming that we are close to a state of equilibrium with a number of vaporization occurrences nearly equal to the number of adsorption occurrences.

Equation [8.8] can be put in a standard formulation:

$$\frac{dF}{dh} = \left\{ \frac{dG_{el.st}}{dh} - \frac{RT}{v_L} \ln \frac{P_{Lab}}{P_0} \right\}$$

At equilibrium (dF=0) and by introducing the disjoining pressure of Derjagin [DER 74] the classic formulation proposed by Bormashenko and Starov [BOR 08, STA 10] is obtained:

$$\Pi(h) = \frac{-dG_{el.st}}{dh} = -\frac{RT}{v_L} \ln \frac{P_{Lab}}{P_0}$$

The term between brackets of equation [8.9] comprises two terms: the first term (electrostatic component) is positive whilst the second one is negative in an imposed situation with a dry atmosphere ($P_{Lab} < P_0$). The compensation between the two terms (dF=0) is obtained for a critical $P_{Lab.}$ value

$$2\kappa_D .G_{el.st}(2h) = -\frac{RT}{v_L} \ln \frac{P_{Lab}^{crit}}{P_0} \qquad [8.10]$$

In a situation where $P_{Lab}^{crit} < P_{Lab} < P_0$, the electrostatic term is dominant in equation [8.9] and $dF / dh < 0$, leading to $dF > 0$ for $dh < 0$. The liquid film resists vaporization.

Conversely, when $P_{Lab} < P_{Lab}^{crit} < P_0$, then $dF/dh > 0$, which implies film vaporization.

A quantification of this effect is possible from equation [8.10] by using typical values of surface charges σ and reasonable values of h, the average depth of the pores.

The other parameters have standard values:

$$T = 293\ K; v_{Liq} = 1.7510^{-5} m^3 / mole; R = 8.314\ JK^{-1} mol^{-1}.$$

Table 8.1 summarizes the results obtained.

$\sigma = 10\mu C/cm^2$, h=10 nm			
[NaCl]	10^{-4}	10^{-6}	10^{-8}
P_{Lab}^{crit} / P_0	0.810	0.684	0.669
$\sigma = 5\mu C/cm^2$, h=10 nm			
[NaCl]	10^{-4}	10^{-6}	10^{-8}
P_{Lab}^{crit} / P_0	0.949	0.909	0.904
$\sigma = 5\mu C/cm^2$, h=30 nm			
[NaCl]	10^{-4}	10^{-6}	10^{-8}
P_{Lab}^{crit} / P_0	0.986	0.927	0.905
$\sigma = 2\mu C/cm^2$, h=70 nm			
[NaCl]	10^{-4}	10^{-6}	10^{-8}
P_{Lab}^{crit} / P_0	1	0.989	0.985

Table 8.1. *Calculated values for the ratio P_{Lab}^{crit} / P_0 using various values of the average depth of the pores, the charge density σ and the salt concentration mole/l*

The stability domain of the prewetting film between P_{Lab}^{crit} and P_0 increases with increasing surface charge ($G_{el.st}$) and decreasing salt concentration (κ_D).

8.2.2.2. *Apparent contact angle*

Very hydrophilic surfaces have been discussed in section 2.3 of Chapter 2 which deals with real surfaces.

Equation [2.8] has been established, which defines a critical angle θ_{crit} such as the super hydrophilic wetting conditions are met:

$$\cos\theta_{crit} = \cos\theta^0_{crit}(1+\frac{G_{LD}}{\gamma_{LV}}) \text{ with } \theta_E \le \theta_{crit} \le \theta^0_{crit} \qquad [2.8]$$

In this equation θ^0_{crit} represents a first critical angle defined by geometrical parameters defining the roughness of the surface. The corrective term G_{LD}/γ_{LV}, which involves the specific energy associated with long distance interactions, will be taken into account by the term $G_{el.st}/\gamma_{LV}$.

Taking into consideration the electrostatic interactions $G_{el.st}$ (having an order of magnitude comparable to γ_{LV}) leads to a substantial reduction of the critical angle θ_{crit}, which hardens the condition $\theta_E \le \theta_{crit} \le \theta^0_{crit}$ which must be verified for a "superhydrophilic" behavior.

In these conditions, the apparent contact angle is given by the Cassie-Baxter equation [2.3]:

$$\cos\theta^* = \phi_s \cos\theta_E + (1-\phi_s) \qquad [2.3]$$

Hence, by considering that θ_E and θ^* arc small:

$$\theta^* = \theta_E\sqrt{\phi_s} \qquad [2.9]$$

A detailed study of the apparent contact angle is given in the publication by Joud *et al.* [JOU 13].

Figure 8.7 summarizes the situation under investigation. The surface charges which exist under the liquid drop are not represented. The electrostatic interactions are quite inactive under the drop, in view of its millimetric dimensions.

A concrete example of the situation presented above is supplied by the study of the TiO_2-SiO_2 [JOU 13] system wettability.

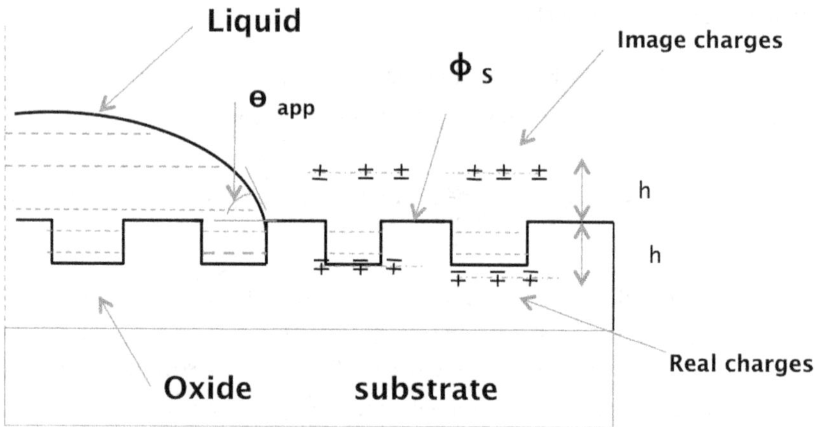

Figure 8.7. *Principle diagram representing a liquid drop on a rough oxide substrate*

The oxide mixture is obtained by a sol–gel process leading to a mixture of TiO_2 anatasis nanoparticles dispersed in an amorphous silica matrix

A 70 nm thick film is deposited (spin-coating) on a silicon substrate, then cleaned and thermally processed at 500°C. It presents a RMS roughness close to 11 nm as well as surface charges (7 $\mu C/cm^2$) measured by AFM in the presence of liquid water at pH 5.6.

The apparent contact angle is measured as a function of the aging time in the laboratory atmosphere on a sample labelled S4-60 corresponding to a molar ratio Si/Si+Ti =60%. The results are shown in Figure 8.8 and compared to those obtained on a pure silica surface.

In the case of pure silica, the evolution of the angle as a function of the aging time presents a classic behavior typical of an increase of the surface carbonaceous contamination with time. In the case of sample S4-60, the apparent angle always remains very small (between 1 and 3 degrees) for exposition times of up to 50 days.

Such a result can only be understood by the persistence of the prewetting film whose stability is linked to the existence of electrostatic interactions.

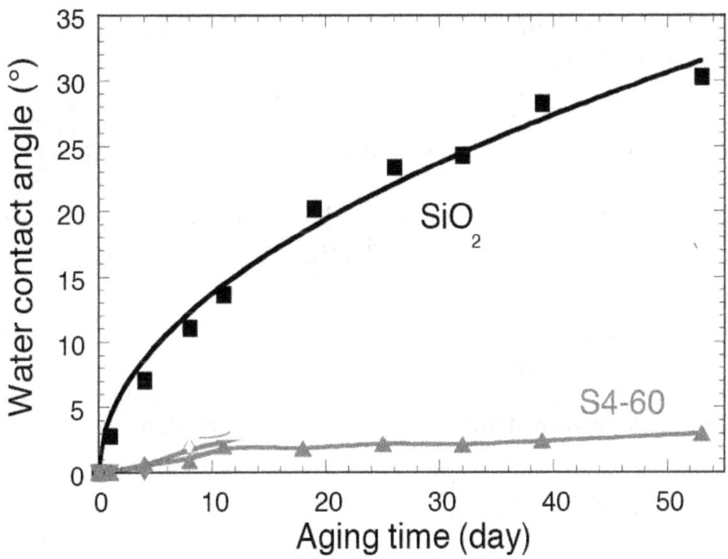

Figure 8.8. *Evolution of the apparent contact angle as a function of the aging time of the sample (pure silica or sample S4-60) in the laboratory atmosphere. The experimental points indicated represent an average of 5 to 7 water droplets.*

Starting from the value of the angle measured on the S4-60 sample kept under atmospheric argon (17–22°), that one assimilates to the equilibrium angle θ_E and supposing a value of the solid fraction ϕ_s of 10%, equation [2.9] allows the calculation of an apparent angle of 5° which is of similar order of magnitude as the experimental results.

The conclusion of this study shows that on an hydrophilic solid, the existence of surface roughness coupled with a disjoining pressure linked to the existence of surface charges can lead to the stability of a prewetting film that is resistant to its vaporization, in some conditions.

Such a result can be applied to semiconductor systems. Similarly, the field of stainless steel corrosion can be concerned. In particular, the stability of a prewetting film containing a standard contaminant like Cl^- can enhance corrosion.

8.3. Modification of the contact angle by an imposed potential (electrowetting)

The modification of the capillary forces at a solid–liquid interface by external electric charges was proposed by Lippmann in 1875 [LIP 75].

The previously established equations [7.23] and [7.28] link the interfacial tension variation to the surface potential Φ_0 :

$$\gamma_{SL} = \gamma_{SL}^{PZC} - \frac{\varepsilon\varepsilon_0}{2\kappa^{-1}}(\Phi_0 - \Phi_{PZC})^2 \quad [8.11]$$

The Gauss theorem allows us to rewrite this relationship:

$$\gamma_{SL} = \gamma_{SL}^{PZC} - \frac{1}{2}\frac{\sigma_0^2\kappa^{-1}}{\varepsilon\varepsilon_0} \quad [8.12]$$

with $\sigma_0 = -\sigma_L = -\varepsilon\varepsilon_0 \dfrac{(\Phi_0 - \Phi_{PZC})}{\kappa^{-1}}$

κ^{-1} is the Debye length which approximates the thickness of the single condenser modeling the solid/liquid interface.

The application of the virtual work theorem in two situations corresponding respectively to a situation where the solid surface possesses a potential Φ_0 and a surface charge density σ_0 and the situation corresponding to the PZC (zero charge and zero potential by convention), leads to the two relationships expressing the free energy variation with area:

$$\frac{dF}{dA} = 0 = \gamma_{SL}^{PZC} - \frac{1}{2}\frac{\sigma_0^2}{\varepsilon\varepsilon_0}\kappa^{-1} - \gamma_{SV} + \gamma_{LV}\cos\theta$$

and

$$\frac{dF}{dA} = 0 = \gamma_{SL}^{PZC} - \gamma_{SV} + \gamma_{LV}\cos\theta_E$$

Hence $\cos\theta = \cos\theta_E + \frac{1}{2\gamma_{LV}}\frac{\varepsilon\varepsilon_0}{\kappa^{-1}}(\Phi_0 - \Phi_{PZC})^2$ [8.13]

Or in a more classic manner

$$\cos\theta = \cos\theta_E + \frac{1}{2\gamma_{LV}}CV^2$$

with C: capacity per unit area of the solid/solution interface and V the surface/solution potential difference.

The angle θ in the presence of charges can only be reduced compared to the angle θ_E corresponding to the equilibrium when there are no charges. Therefore, to have a significant amplitude of variation of the wetting angle, it is necessary to work on an insulating oxide surface covered with a hydrophilic film presenting a large θ_E angle. The surface layer under consideration must cover a metal electrode required for the electric contact. The capacity of the coated system is mainly given by the insulating dielectric capacity (capacities in series).

In order of magnitude $V \approx (2\gamma_{LV} / C)^{1/2} \approx (2\gamma_{LV} d / \varepsilon\varepsilon_0)^{1/2}$

with d and ε the thickness and the relative permittivity of the insulator.

For a coated system, $d \approx 10\mu m$ and $\varepsilon \approx 10$.

For the metal/solution system, $d \approx 1nm$ and $\varepsilon \approx 80$.

The tension ratio is 10^3, i.e. the tensions to be applied are several tens of volts for the coated system compared to several tens of millivolts for a metal-solution system.

Figures 8.9 and 8.10 depict the two situations.

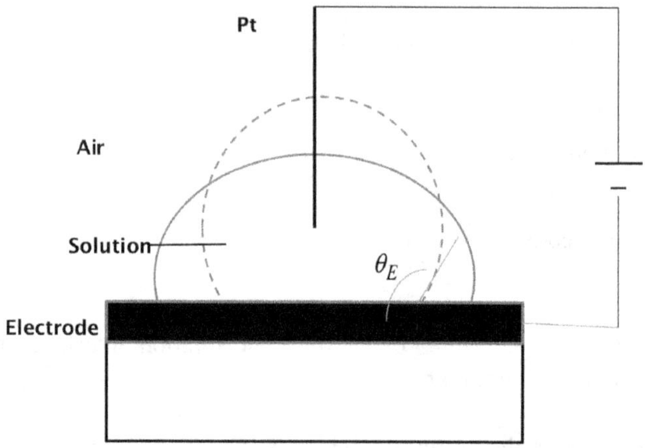

Figure 8.9. *Metal electrode/solution system. The electric contact with the conducting solution is carried out by a platinum electrode. Dotted line: drop shape in the absence of charges; continuous line: drop shape in the presence of charges*

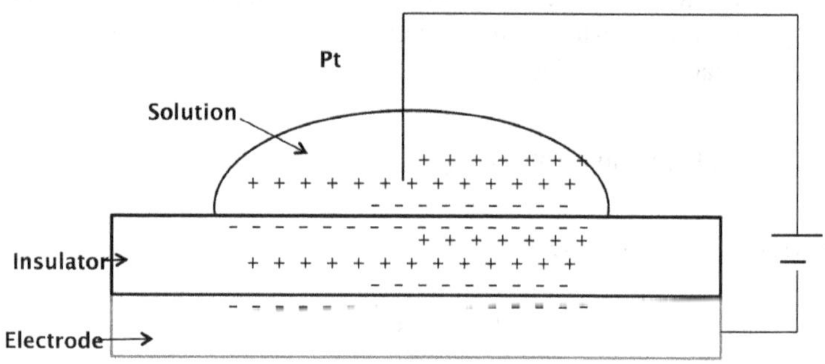

Figure 8.10. *Equilibrium configuration, in the presence of charges, when an insulating layer is inserted between the liquid solution and the metal electrode*

The application of electro-wettability for about 15 years is linked to the development of microfabrication techniques, enabling the implementation of microfluidic systems (micro electromechanics system: MEMS and Laboratory on chip: Lab. on chip)

A number of publications review this aspect of electro-wettability and of its applications [VAL 96, VER 99, LEE 02, MOO 02, VAL 99, MAL 12].

The evolution of the contact angle with the imposed potential roughly follows the parabolic law given by equation [8.13]. However, for large values of the potential, a saturation phenomenon appears, linked to the maximum quantity of charges acceptable at the solid/liquid interface, or even to the existence of charges inserted in the insulator.

This latter effect was taken into consideration by Verheijen and Prins [VER 99] leading to a correction of equation [8.13]:

$$cos\theta - \cos\theta_0 = \frac{1}{2\gamma_{LV}} C(V - V_T)^2$$

For the charge density inserted in the depth d_1, $\sigma_T = \frac{\varepsilon\varepsilon_0}{d_1} V_T$; and for the surface charge density $\sigma_L = \frac{\varepsilon\varepsilon_0}{d}(V - V_T)$ for the dielectric of total thickness d.

The electrowetting force is reduced; it becomes proportional to the square of the difference between the imposed potential V and the potential associated with the implanted charges V_T.

These results are illustrated in Figure 8.11 taken from the work by Verheijen and Prins [VER 99].

The system is made up of a conducting aluminum layer 100 nm thick. The insulator coating comprises a pyralene layer ($d = 10\ \mu m\ \ \varepsilon = 2.65$). The coating is covered by a very thin Teflon hydrophobic layer filled with silicon oil. The electrode in contact with the solution is a platinum wire. Finally, the liquid droplet (10 µl) consists of an aqueous solution loaded with salt (KCl or K_2SO_4 0.1M).

The measurement of the system capacity is achieved by superimposing an AC voltage (5 V, 700 Hz) on the imposed DC voltage. The contact angle is therefore deduced from the physical parameters of the drop (contact surface, volume and shape) and from the insulator characteristics (thickness and permittivity).

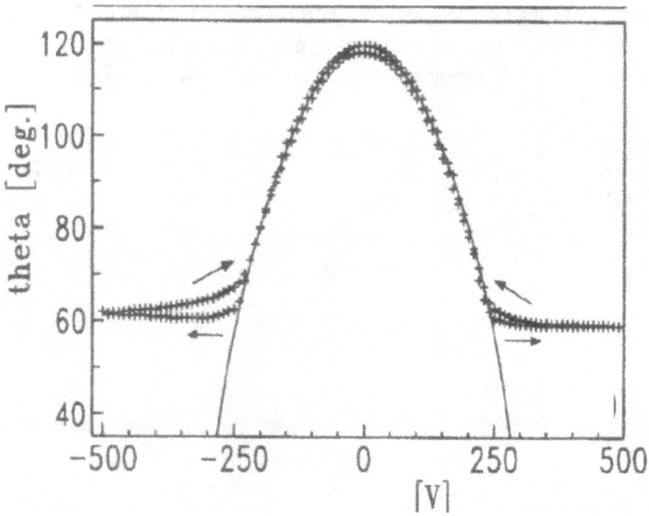

Figure 8.11. *Evolution of the contact angle as a function of the DC voltage imposed to the platinum electrode /aqueous solution/ insulating pyralene/aluminum electrode system [VER 99]. The continuous curve gives the evolution of the contact angle calculated from equation [8.13]; copyright 1999 with permission of the American Chemical Society*

Various settings have been proposed for moving the liquid drop set on appropriate substrates by applying an AC or DC electric voltage.

The implementation of continuous electrodes placed on the back of an insulator in direct contact with the fluid allows setting in motion the totality of the fluid as shown in Figure 8.12 taken from the work by Lee *et al.* [LEE 02].

Conversely, the use of discontinuous electrodes covered with insulators which can be sequentially activated (EWOD electrowetting on dielectric) allows setting a liquid drop in motion [LEE 02, BAN 11].

Figure 8.13, taken from the work by Lee *et al.* [LEE 02], indicates a possible configuration.

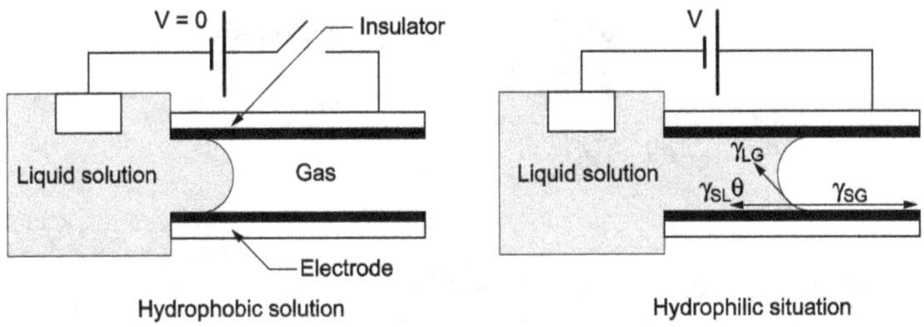

Figure 8.12. *Continuous electrode configuration enabling moving the totality of the fluid [LEE 02], copyright 2002 with permission of Elsevier*

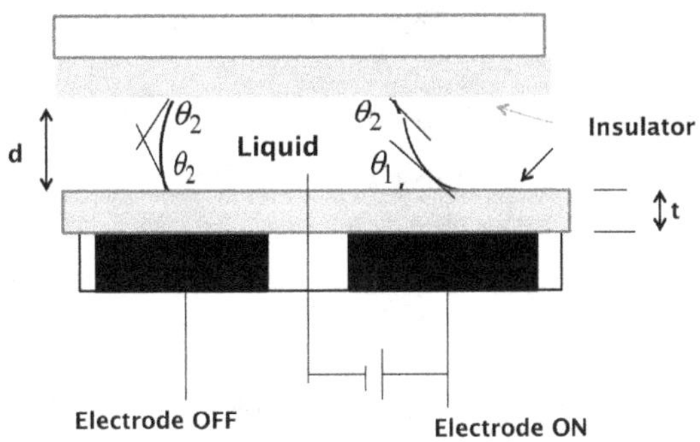

Figure 8.13. *Discontinuous electrode configuration enabling moving a liquid drop [LEE 02], copyright 2002 with permission of Elsevier*

The Laplace equation can be applied to a liquid–gas interface which presents a finite radius of curvature on one side and an infinite one on the other side to calculate the pressure variation on either side of the interface: $\Delta P = \gamma_{LG} \dfrac{1}{R}$.

This equation, applied successively to the left and to the right sides of Figure 8.13 leads to:

$$[P_L - P_G]_{\text{left}} = -\frac{\gamma_{LG}}{d}(\cos\theta_2 + \cos\theta_2)$$

$$[P_L - P_G]_{\text{right}} = -\frac{\gamma_{LG}}{d}(\cos\theta_1 + \cos\theta_2)$$

Hence by difference a driving pressure in the liquid can be written using equation [8.13]:

$$\Delta P = (P_L)_{\text{left}} - (P_L)_{\text{right}}$$

$$\Delta P = \frac{\gamma_{LG}}{d}(\cos\theta_1 - \cos\theta_2) = \frac{1}{d}(\frac{1}{2}CV^2) = \frac{\varepsilon\varepsilon_0 V^2}{2dt}$$

where d is the distance between the 2 insulating layers and t is the insulator thickness.

The largest movement speed is obtained in the configuration indicated in Figure 8.14 without any upper electrode which induces braking [BAN 11].

The surface driving force $F_{surf} = 2r\gamma_{LV}(\cos\theta_1 - \cos\theta_2)$ is assimilated to the viscous force. It is written as the product of the contact surface πr^2 by the shearing stress:

$F_{\text{shearing}} = \pi r^2 \eta \frac{U}{H}$ with U the movement speed η the liquid viscosity and H the height corresponding to the maximum speed in the drop.

Banerjee *et al.*'s results [BAN 11] show that a speed up to 26 cm/s can be obtained with an insulator like SiO_2 (0.3 μm) and an activation of 110 V.

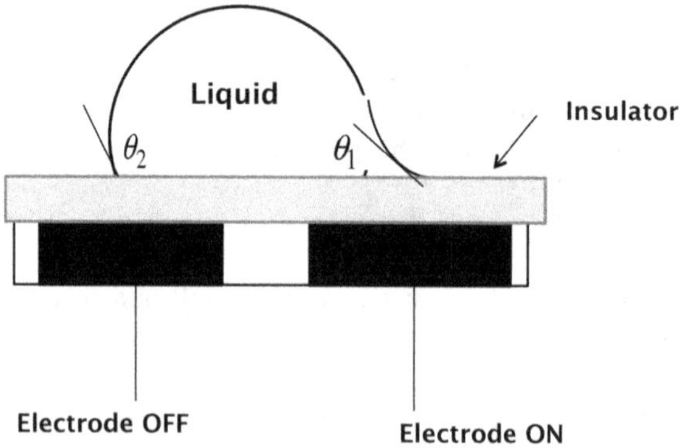

Figure 8.14. *Flat configuration with discontinuous electrodes [BAN 11], copyright 2011 with permission of Elsevier*

This configuration is the most favorable for liquid evaporation. However, the evaporation time of a liquid drop of several μL in a dry atmosphere without convection is evaluated to be 25–30 min [BAN 11], which leaves reasonable time for the transfer of the drop.

The activation of a liquid or of liquid drops in the implementation of microfluidics (micro electromechanics system: MEMS and Laboratory on chip: Lab. on chip) currently represents a field of significant development.

Another axis of significant development, supported in particular by the company Varioptics (http://varioptic.com), involves the variable focal lenses.

The principle consists of modulating the curvature of an interface between two immiscible fluids (transparent and of the same density) such as water/oil by an electric voltage action.

The curvature change induced by the voltage enables us to obtain various focals without mechanical movement in the device.

More traditional applications can also be considered. For example, some thermal converters can be driven due to the action of an electric voltage on a metal wall.

Conclusion

The physical chemistry of surfaces and their acid-base properties is a huge topic. In this work, far from being exhaustive, only those issues which are important regarding their consequences on the wettability properties or on acid-base properties have been considered. To a large extent, the examples presented are obtained from the authors' research work.

Following assessing the case of ideal surfaces and surfaces with defects, the various components involved in the expression of the solid–liquid or liquid–vapor interfacial tensions are introduced with the help of an in-depth analysis of the literature. The experimental determinations of surface tension through wettability are then presented and illustrated by several examples using the various models which are proposed in the literature.

A review of the different experimental methods allowing to characterize the acid-base properties of the surfaces is then proposed, with an emphasis on those methods mostly used in the case of flat surfaces.

Electrocapillarity is then the topic of a detailed chapter, which highlights the importance of electric charges on the solid–liquid interfacial tension. A new method is proposed for the evaluation of the maximum charge at the solid–liquid interface from measurements of contact angles as a function of the pH of the liquid solution.

Finally, the last chapter presents several applications of electrocapillarity and ends with the important topic of electrowetting, that is to say the modification of the contact angle by an imposed electric voltage.

Acknowledgments

Dr. Dupuy, Research Engineer at the CEA-Leti, and Dr A. Passeronne, Research Director at the C.N.R. Genova, are gratefully thanked for their careful reading of the manuscript and their constructive remarks on its formatting. Special thanks go to the authors' families for their patience and their encouragement during the preparation of this book.

Bibliography

[ADA 90] ADAMSON A.W., *Physical Chemistry of Surfaces*, 5th ed., John Wiley & Sons, New York, 1990.

[BAL 00] BALARD H., BRENDLE E., PAPIRER E., "Determination of the acid-base properties of solid surfaces using inverse gas chromatography: advantages and limitations", in MITTAL K.L. (ed.), *Acid-Base Interactions: Relevance to Adhesion Science and Technology*, VSP, pp. 299–316, 2000.

[BAN 11] BANERJEE A.N., QIAN S., JOO S.W., "High-speed droplet actuation on single plate electrode arrays", *Journal of Colloid and Interface Science,* vol. 362, p. 567, 2011.

[BAR 90] BAROUX B., "Passivité et rupture de la passivité des aciers inoxydables", in LACOMBE P., BAROUX B., BÉRANGER G. (eds), *Les aciers inoxydables*, Les Editions de Physique, Les Ulis, pp 161–182, 1990.

[BAR 94] BAROUX B., GORSE D., "Physique des films passives", in DABOSI F., BÉRANGER G., BAROUX B. (eds), *Corrosion localisée*, Les Editions de Physique, pp. 91–147, 1994.

[BEN 83] BENARD J. (ed.), "E-hondros in adsorption on metal surfaces", *Equilibrium Adsorption and Surface Segregation, Studies in Surface Science and Catalysis*, Elsevier Scientific Publishing Company, New York, vol. 13, 1983.

[BIC 02] BICO J., THIELE U., QUÉRÉ D., "Wetting of textured surfaces", *Colloids and Surfaces: A Physicochemical and Engineering Aspects*, vol. 206, pp. 41–46, 2002.

[BIN 86] BINNIG G., QUATE C.F., GERBER C., "Atomic force microscope", *Physical Review Letters*, vol. 56, pp. 930–933, 1986.

[BOC 70] BOCKRIS J.O.M., REDDY A.K.N., *Modern Electrochemistry T2*, Plenum Press, New York, 1970.

[BOL 68] BOLGER J.C., MICHAELS A.S., "Acid base interactions between oxide surface and polar organic compound", in WEISS P., CHEEVER G. (eds), *Interface Conversion for Polymer Coatings*, Elsevier, New York, pp. 3–60, 1968.

[BOR 07] BORMASHENKO E., BORMASHENKO Y., STEIN T. *et al.*, "Cassie-Baxter wetting hypothesis and Cassie-Wenzel capillarity induced wetting transition", *Journal of Colloid and Interface Science*, vol. 311, pp. 212–216, 2007.

[BOR 08] BORMASHENKO E., PROGERB R., STEIN T., "Characterization of rough surface with vibrated drops", *Physical Chemistry and Chemical Physics*, vol. 10, p. 4056, 2008.

[BOR 13] BORMASHENKO E., STAROV V.M., "Impact of surface forces on wetting of hierarchical surfaces and contact angle hysteresis", *Colloid and Polymer Science*, vol. 291, p. 343, 2013.

[BOS 13] BOSCOBOINIK J.A., YU X., EMMEZ E. *et al.*, "Interaction of probe molecules with bridging hydroxyls of two-dimensional zeolites: a surface science approach", *Journal of Physical Chemistry C*, vol. 117, pp. 13547–13556, 2013.

[BOU 95] BOULANGÉ-PETERMANN L., DOREN A., BAROUX B. *et al.*, "Zeta-potential measurements on passive metals", *Journal of Colloid and Interface Science*, vol. 171, pp. 179–186, 1995.

[BRE 97] BRENDLE E., PAPIRER E., "A new topological index for molecular probes used in inverse gas chromatography .2. Application for the evaluation of the solid surface specific interaction potential", *Journal of Colloid and Interface Science*, vol. 194, p. 217, 1997.

[BRU 95] BRUEL M., "Silicon on insulator material technology", *Electronic Letter*, vol. 31, no. 14, p. 1201, 1995.

[BUT 78] BUTLER M.A., GINLEY D.S., "Prediction of flatband potentials at semiconductor-electrolyte interfaces from atomic electronegativities", *Journal of the Electrochemical Society*, vol. 125, p. 228, 1978.

[CAS 44] CASSIE A.B.D., BAXTER S., *Transaction of the Faraday Society*, vol. 40, p. 546, 1944.

[CAS 91a] CASAMASSIMA M., DARQUE-CERETTI E., DELAMARE F. *et al.*, "XPS characterization od surface electron-donor acceptor properties of aluminum and silicon oxides", *Compte Rendus de l'Académie des Sciences*, vol. 313, no. 2, pp. 25–30, 1991.

[CAS 91b] CASAMASSIMA M., DARQUE-CERETTI E., ETCHEBERRY A. *et al.*, "Acid-base behavior of aluminum and silicon oxides – a combination of 2 approaches – XPS and Lewis acido-basicity – rest potential and Bronsted acido-basicity", *Applied Surface Science*, vol. 52, pp. 205–213, 1991.

[CHA 13] CHAPMAN D.L., "A contribution to the theory of electrocapillarity", *Philosophical Magazine*, vol. 25, pp. 475–481, 1913.

[CHE 00] CHEHIMI M.M., DELAMAR M., KURDI J. *et al.*, "Characterisation of acid-base properties of polymer surfaces by XPS", in MITTAL K.L. (ed.), *Acid-Base Interactions: Relevance to Adhesion Science and Technology*, VSP, Utrecht, pp. 275–298, 2000.

[CHE 02] CHEVALLIER I., Influence des interactions acido-basiques sur l'encrassement et la biocontamination de supports métalliques oxydés, PhD Thesis, University Paris XI Orsay, 2002.

[DE 03] DE GENNES P.-G., BROCHARD-WYART F., QUÉRÉ D. *et al.*, *Perles et ondes*, Belin, Paris, 2002; *Capillarity and Wetting Phenomena*, Springer, Berlin, 2003.

[DEB 01] DEBONTRIDDER F., Influence de l'acidité de surface sur les mécanismes précurseurs de formation d'une interphase dans les assemblages epoxy-aluminium, PhD Thesis, Paris XI, Paris, 2001.

[DEF 66] DEFAY R., PRIGOGINE I., *Surface Tension and Adsorption*, Longmans, Green & Company, London, 1966.

[DEF 78] DEFOSSÉ C., CANESSON P., ROUXHET P.G. *et al.*, "Surface characterization of silica-aluminas by photoelectron-spectroscopy", *Journal of Catalysis*, vol. 51, pp. 269–277, 1978.

[DEL 07] DELGADO A.V., GONZALEZ-CABALLERO F., HUNTER R.J. *et al.*, "Measurement and interpretation of electrokinetic phenomena", *Journal of Colloid and Interface Science*, vol. 309, pp. 194–224, 2007.

[DER 34] DERJAGIN B.V., *Kolloid Zhournal*, vol. 69, pp. 155–164, 1934.

[DER 41] DERJAGIN B.V., LANDAU L., *Acta Physicochimii URSS*, vol. 14, pp. 633–662, 1941.

[DER 74] DERJAGIN B.V., CHURAEV N.V., *Journal of Colloid and Interface Science*, vol. 49, p. 249, 1974.

[DER 75] DERJAGUIN B.V., MULLER V.M., TOPOROV Y.P., "Effect of contact deformations on adhesion of particles", *Journal of Colloid and Interface Science*, vol. 53, p. 314, 1975.

[DON 89] DONG S., BRENDLÉ M., DONNET J.B., "Study of solid-surface polarity by inverse gas-chromatography at infinite dilution", *Chromatographia*, vol. 28, p. 85, 1989.

[DON 91] DONNET J.B., PARK S.J., BALARD H., "Evaluation of specific interactions of solid-surfaces by inverse gas-chromatography – a new approach based on polarizability of the probes," *Chromatographia*, vol. 31, p. 434, 1991.

[DOR 80] DORRIS G.M., GRAY D.G., "Adsorption of normal-alkanes at zero surface coverage on cellulose paper and wood fibers", *Journal of Colloid and Interface Science*, vol. 77, pp. 353–362, 1980.

[DRA 65] DRAGO R.S., WAYLAND B., "A double-scale equation for correlating enthalpies of Lewis acide-base interactions", *Journal of the American Chemical Society*, vol. 87, p. 3571, 1965.

[DRA 77] DRAGO R.S., VOGEL G.C., NEEDHAM T.E., "A four parameter equation for predicting enthalpies of adduct formation", *Journal of the American Chemical Society*, vol. 93, p. 3203, 1977.

[DRE 11] DRELICH J., CHIBOWSKI E., DESHENG D. *et al.*, "Hydrophilic and superhydrophilic surfaces and materials", *Soft Matter*, vol. 7, pp. 9804–9828, 2011.

[DUB 00] DUBOIS P.-E., JOUD J.-C., *Le Vide: Science, Technique et Applications*, vol. 296, p. 476, 2000.

[DUB 00] DUBOIS P.E., Influence de l'acido-basicité de surface sur les propriétés d'adhésion et de durabilité d'oxydes métalliques. Application au système aciers inoxydables-époxydes, PhD Thesis, INP Grenoble, 2000.

[FOW 62] FOWKES F.M., *Journal of Physical Chemistry*, vol. 66, p. 382, 1962.

[FOW 64] FOWKES F.M., "Attractive forces at interfaces", *Industrial and Engineering Chemistry*, vol. 56, p. 40, 1964.

[FOW 70] FOWKES F.M., *Industrial and Engineering Chemistry*, vol. 40, pp. 56–12, 1964; *Chemistry and Physics of Interfaces*, American Chemical Society, Washington D.C., 1970.

[FOW 78] FOWKES F.M., MOSTAFA M.A., "Acid-base interactions in polymer adsorption", *Industrial and Engineering Chemistry Product Research and Development*, vol. 17, p. 3, 1978.

[FOW 80] FOWKES F.M., MC CARTHY D.C., MOOSTAFA M.A., "Contact angles and the equilibrium spreading pressure of liquids on hydrophobic solids", *Journal of Colloid and Interface Science*, vol. 78, p. 200, 1980.

[FRA 59] FRANCIS S.A., ELLISON A.H., "Infrared spectra of monolayers on metal mirrors", *Journal of the Optical Society of America*, vol. 49, pp. 131–137, 1959.

[GIL 64] GILLIS P.P., GILMAN J.J., *Journal of Applied Physics*, vol. 35, p. 647, 1964.

[GOO 64] GOOD R.J., *Contact Angle Wettability and Adhesion*, Advances in Chemistry Series, American Chemical Society, Washington D.C., N 43, p. 74, 1964.

[GOO 91] GOOD R.J., CHAUDHURY M.K., VAN OSS C.J., in LEE L.H. (ed.), *Fundamentals of Adhesion*, Plenum Press, New York, p. 153, 1991.

[GOS 99] GOSELE U., Tong Q., Schumacher A. *et al.*, "Wafer bonding for microsystems technologies", *Sensors and Actuators*, vol. 74, p. 161, 1999.

[GOU 09] GOUY L.G., "On the constitution of the electric charge at the surface of an electrolyte", *Comptes rendus hebdomadaires des séances de l'Académie des Sciences*, Paris, vol. 149, pp. 654–657, 1909.

[GRA 47] GRAHAME D.C., "The electrical double layer and the theory of electrocapillarity", *Chemical Reviews*, vol. 41, pp. 441–501, 1947.

[GRO 13] GROSSO S., Influence de la composition du film passif d'aciers inoxydables ferritiques sur la biocontamination, Master, INP Grenoble, 2013.

[GUT 78] GUTMANN V., *The Donor-Acceptor Approach to Molecular Interactions*, Plenum, New York, 1978.

[HAM 37] HAMAKER H.C., "The London Van Der Waals attraction between sphericale particules", *Physica*, vol. 4, pp. 1058–1072, 1937.

[HEL 53] HELMHOLTZ H., "Ueber einige Gesetze der Vertheilung elektrischer Ströme in körperlichen Leitern, mit Anwendung auf die thierisch-elektrischen Versuche", *Annalen der Physik und Chemie*, vol. 165, pp. 211–233, 1853.

[HOU 08] HOUMARD M., RIASSETO D., ROUSSEL F. *et al.*, "Enhanced persistence of natural superhydrophilicity in TiO2-SiO2 composite thin films deposited via a sol-gel route", *Surface Science*, vol. 3364, pp. 602–621, 2008.

[HOU 11] HOUMARD M., BERTHOME G., JOUD J.-C. *et al.*, "Enhanced cleanability of superhydrophilic TiO2-SiO2 composite surfaces prepared via a sol-gel route", *Surface Science*, vol. 605, pp. 456–462, 2011.

[HUD 73] HUDSON H., HUDSON R.F., *Angewandte Chemie*, vol. 36, pp. 12–11, 1973.

[ISR 11] ISRAELACHVILI J.N., *Intermolecular and Surface Forces*, 3rd ed., Academic Press, London, 2011.

[ISR 89] ISRAELACHVILI J.N., *Intermolecular and Surfaces Forces: Applications to Colloidal and Biological Systems*, Academic Press, New York, 1989.

[ISR 91] ISRAELACHVILI J.N., *Intermolecular and Surface Forces*, 2nd ed., Academic Press, London, 1991.

[JOH 64] JOHNSON R.E., DETTRE R.H., "Contact angle, wettability and adhesion", *Advances in Chemistry Series*, vol. 43, p. 112, 1964.

[JOH 71] JOHNSON K.L., KENDALL K., ROBERTS A.D., "Surface energy and contact of eleastic solides", *Proceedings of the Royal Society of London Series A-Mathematical and Physical Sciences*, vol. A 324, p. 301, 1971.

[JOL 94] JOLIVET J.P., *De la solution à l'oxyde*, InterEditions/CNRS Editions, Paris, 1994.

[JOU 98] JOUD J.C., VITTOZ C., MANTEL M., "Wettability of metallic oxides – application to stainless steel acide-base properties" *The Journal of Adhesion*, vol. 67, p. 347, 1998.

[JOU 13] JOUD J.-C., HOUMARD M., BERTHOMÉ G., "Surface charges of oxides and wettability: application to TiO2-SiO2 composite films", *Applied Surface Science*, vol. 287, pp. 37–45, 2013.

[JOY 91] JOYCE S.A., HOUSTON J.E., "A new force sensor incorporating force-feedback control for interfacial force microscopy", *Review of Scientific Instruments*, vol. 62, p. 710, 1991.

[KAE 70] KAELBLEET D.H., UY K.C., *Journal of Adhesion*, vol. 2, p. 50, 1970.

[KIN 87] KINLOCH A.J., *Adhesion and Adhesive*, Chapman and Hall, London, 1987.

[KIT 76] KITAURA K., MOROKUMA K., "New energy decomposition scheme for molecular interactions within Hartree Fock approximation", *International Journal of Quantum Chemistry*, vol. 10, p. 325, 1976.

[KLO 67] KLOPMAN G., HUDSON, *Theoretical Chimica Acta*, vol. 8, p. 165, 1967.

[KLO 68] KLOPMAN G., "Chemical reactivity and concept of charge-and frontier-controlled reactions", *Journal of the American Chemical Society*, vol. 90, p. 543, 1968.

[KUR 92] KURBATOV G., DARQUE-CERETI E., "Characterization of hydroxylated oxide film on iron surfaces and its acid-base properties using XPS", *Surface and Interface Analysis*, vol. 18, pp. 811–820, 1992.

[KUR 93] KURBATOV G., DARQUE-CERETTI E., AUCOUTURIER M., "XPS study of hydroxylated oxide layer formation on pure iron and their acid-base properties", *Surface and Interface Analysis*, vol. 20, pp. 402–406, 1993.

[LAB 84] LABIB M.E., WILLIAMS R., "The use of zeta-potential measurements in organic-solvents to determine the donor-acceptor properties of solid surfaces", *Journal of Colloid and Interface Science*, vol. 97, p. 356, 1984.

[LAN 38] LANGMUIR I., *Journal of Chemical Physics*, vol. 6, p. 893, 1938.

[LAN 93] LANDOLT D., *Corrosion et Chimie de surfaces des métaux*, Presses Polytechniques et Universitaires Romandes, Lausanne, 1993.

[LAS 86] LASKY B., "Wafer bounding for Silicon-on-insulator technologies", *Applied Physics Letters*, vol. 48, p. 78, 1986.

[LAV 89] LAVIELLE L., "The wettability", *Annales de Physique*, vol. 14, pp. 1–48, 1989.

[LEE 89] LEE L.H., in LEE (ed.), *New Trends in Physics and Physical Chemistry of Polymers*, Plenum Press, New York, p. 185, 1989.

[LEE 93] LEE L.H., "Roles of molecular interactions in adhesion, adsorption, contact angle and wettability", in MITTAL K.L. (ed.), *Contact Angle, Wettability and Adhesion*, VSP, Utrecht, the Netherlands, pp. 46–96, 1993.

[LEE 02] LEE J., MOON H., FOWLER J. *et al.*, "Electrowetting and electrowetting on dielectric for microscale handling", *Sensors and Actuators*, vol. A95, p. 259, 2002.

[LEW 23] LEWIS G.N., *Valence and Structure of Atoms and Molecules*, The Chemical Catalog Co., New York, 1923.

[LIN 93] LIN X.Y., CREUZET F., ARRIBART H., "Atomic-force microscopy for local characterization of surface acid-base properties", *Journal of Physical Chemistry*, vol. 97, pp. 7272–7276, 1993.

[LIP 75] LIPPMANN G., *Annales de Chimie et de Physique*, vol. 5, p. 494, 1875.

[LOP 97] LOPEZ S., Acido-basicité et hydroxylation/hydratation de surfaces d'aluminium, PhD Thesis, INP Grenoble, 1997.

[LOP 98] LOPEZ S., PETIT J.P., DUNLOP H.M. *et al.*, "Acid-base properties of passive films on aluminum I. A photoelectrochemical study", *Journal of the Electrochemical Society*, vol. 145, pp. 823–829, 1998.

[MAL 12] MALK R., THEISEN J., FOUILLET Y. *et al.*, "EWOD-driven stirring in Lab-on-a-chips: dependance on the electrodes' geometry", *Microelectronic Engineering*, vol. 97, p. 306, 2012.

[MAS 88] MASZARA W.P., GOETZ G., CAVIGLIA A. *et al.*, *J. Appl. Phys.*, vol. 64, p. 4943, 1988.

[MER 08] MERCIER D., ROUCHAUD J.-C., BARTHÉS-LABROUSSE M.-G., "Interaction of amines with native aluminium oxide layers in non-aqueous environment: application to the understanding of the formation of epoxy-amine/metal interphases", *Applied Surface Science*, vol. 254, pp. 6495–6503, 2008.

[MER 99] MERCIER V., Etude des mécanismes de formation de l'interphase dans les systèmes époxy-amine/aluminium, PhD Thesis, INP Grenoble, 1999.

[MOO 02] MOON H., CHO S.K., GARRELL R.L. *et al.*, "Low voltage electrowetting -on-dielectric", *Journal of Applied Physics*, vol. 4080, pp. 92–97, 2002.

[MOR 80] MOROKUMA K., KITAURA, in RATAJCZAK H., ORVILLE-THOMAS W. (eds), *Molecular Interactions*, John Wiley, New York, 1980.

[MOR 07] MORICEAU H., RIEUTORD F., Morales C. *et al.*, "Direct water bonding for nanostructure preparations", *Solid State Phenomena*, vol. 29, p. 121, 2007.

[MOR 80] MORRISON S.R., *Electrochemistry at Semiconductor and Oxidized Metal Electrodes*, Plenum Press, New York, 1980.

[MUK 95] MUKHOPADHYAY P., SCHREIBER H.P., "Aspects of acid-base interactions and use of inverse gas-chromatography", *Colloids and Surfaces A-Physicochemical and Engineering Aspects*, vol. 100, p. 47, 1995.

[MUL 51] MULLIKEN R.S., "Lewis acids and bases and molecular complexes", *Journal of Chemical Physics*, vol. 19, p. 514, 1951.

[MUL 88a] MULLINS W.M., AVERBACH B.L., "Bias-reference X-ray photoelectron-spectroscopy of sapphire and yttrium aluminum garnet crystals", *Surface Science*, vol. 206, pp. 29–40, 1988.

[MUL 88b] MULLINS W.M., AVERBACH B.L., "Surface properties of silicon and aluminum-oxide powders", *Surface Science*, vol. 206, pp. 41–51, 1988.

[MUL 88c] MULLINS W.M., AVERBACH B.L., "The electronic-structure of anodized and etched aluminum-alloy surfaces", *Surface Science*, vol. 206, pp. 52–60, 1988.

[NOL 65] NOLLER C.R., in SAUNDERS W. (ed.), *Chemistry of Organic Compounds*, Philadelphia, pp. 983–991, 1965.

[OWE 69] OWENS D.K., WENDT R.C., *Journal of Applied Polymer Science*, vol. 13, p. 1740, 1969.

[PAR 65] PARKS G.A., "The isoelectric points of solid oxides, solid hydroxides and aqueous hydrocomplex systems", *Chemical Review*, vol. 65, p. 177, 1965.

[PAR 10] PARRY V., APPERT E., JOUD J.-C., "Characterisation of wettability in gas diffusion layer in proton exchange membrane fuell cells", *Applied Surface Science*, vol. 256, pp. 2474–2478, 2010.

[PAR 12] PARRY V., BERTHOMÉ G., JOUD J.-C., "Wetting properties of gas diffusion layers: application of the Cassie-Baxter and Wenzel equations", *Applied Surface Science*, vol. 5619, pp. 258–315, 2012.

[PLE 80] PLESKOV Y.V., in BOCKRIS J.O.M., CONWAY B.E., in YEAGER E. (ed.), *Comprehensive Treatise of Electrochemistry*, Plenum Press, New York, vol. 1, p. 291, 1980.

[PLO 99] PLOBL A., KRAUTER G., "Wafer direct bonding: tailoring adhesion between brittle materials", *Material Science and Engineering*, vol. R25, p. 1, 1999.

[RAU 14] RAUER C., Collages de silicium et d'oxydes de silicium: mécanismes, PhD thesis, University of Grenoble, July 2014.

[RED 11] REDDI RANI P., RAMANAIAH S., REDDY K.S., "Lewis acid-base properties of cellulose acetate butyrate by inverse gas chromatography", *Surface and Interface Analysis*, vol. 43, pp. 683–688, 2011.

[RIE 01] RIEUTORD F., EYMERY J., FOURMEL F. *et al.*, "High-energy X-ray reflectivity of burried interfaces created by wafer bonding", *Physical Review B*, vol. 63, p. 1254, 2001.

[SAI 83] SAINT-FLOUR C., PAPIRER E., "Gas-solid chromatography - a quick method of estimating surface free-energy variations induced by the treatment of short glass-fibers", *Journal of Colloid and Interface Science*, vol. 91, p. 69, 1983.

[SAT 89] SATO N., "Towards a more fundamental understanding of corrosion processes", *Corrosion*, vol. 45, pp. 354–368, 1989.

[SCH 77] SCHULTZ J., TSUTSUMI K., DONNET J.B., "Surface properties of high-energy solids", *Journal of Colloid and Interface Science*, vol. 59, p. 272, 1977.

[SCH 87] SCHULTZ J., LAVIELLE L., MARTIN C., "The role of the interface in carbon-fiber epoxy composites", *Journal of Adhesion*, vol. 23, p. 45, 1987.

[SHI 86] SHIMBO M., FURUKAWA K., FUKUDA K. *et al.*, "Silicon to silicon direct bonding", *Journal of Applied Physics*, vol. 60, p. 2987, 1986.

[SHI 96] SHIBUICHI A., ONDA T., SATOH N. *et al.*, "Super water- repellent surfaces resulting from fractal structure", *Journal of Physical Chemistry*, vol. 100, p. 19512, 1996.

[SHU 50] SHUTTLEWORTH R., *Proceedings of the Physical Society*, vol. 63A, p. 444, 1950.

[SIE 67] SIEGBAHN K., NORDLING C., FAHLMAN A. *et al.*, "ESCA – atomic, molecular and solid state structure studied by means of electron spectroscopy", *Nova Acta Regiae, Soc. Sci. Upsaliensis*, vol. 20, no.4, 1967.

[SIM 87] SIMMONS G.W., BEARD B.C., "Characterisation of acid-base properties of the hydrated oxides on iron and titanium metal-surfaces", *Journal of Physical Chemistry*, vol. 91, pp. 1143–1148, 1987.

[STA 10] STAROV V.M., "Surface forces action in a vicinity of three phase contact line and other current problems in kinetic of wetting and spreading", *Advances in Colloid and Interface Science*, vol. 161, p. 139, 2010.

[STE 24] STERN O., "The theory of the electrolytic double shift", *Zeitschrift für Elektrochemie und Angewandte Physikalische Chemie*, vol. 30, pp. 508–516, 1924.

[STE 89] STENGL R., AHN K., MLL T. *et al.* "Tunneling structures fabricated by Silicon-wafer direct bonding", *Japanese Journal of Applied Physics*, vol. 10, no. 28, p. 1735, 1989.

[STO 80] STOL R.J., DE BRUYN P.N., "Thermodynamic stabilization of colloids", *Journal of Colloid and Interface Science*, vol. 185, pp. 75–81, 1980.

[THO 95] THOMAS R.C., HOUSTON J.E., CROOKS R.M. *et al.*, "Probing adhesion forces at the molecular scale", *Journal of the American Chemical Society*, vol. 117, pp. 3830–3834, 1995.

[VAL 96] VALLET M., BERGE B., VOVELLE L., "Electrowetting of water and aqueous solutions on PET insulating films", *Polymer*, vol. 2465, pp. 37–42, 1996.

[VAL 99] VALLET M., VALLADE M., BERGE B., "Limiting phenomena for the spreading of water on polymer films by electrowetting", *European Physical Journal*, vol. B-11, p. 583, 1999.

[VAN 06] VAN OSS C.J., *Interfacial Forces in Aqueous Media*, 2nd ed., Taylor and Francis, 2006.

[VAN 80] VAN WAGENEN R.A., ANDRADE J.D., "Flat-plate streaming potential investigations – hydrodynamics and electrokinetic equivalency", *Journal of Colloid and Interface Science*, vol. 76, pp. 305–314, 1980.

[VAN 86] VAN OSS C.J., GOOD R.J., CHAUDHURY M.K., "The role of Vanderwaals forces and hydrogen-bonds in hydrophobic interactions between bio-polymers and low-energy surfaces", *Journal of Colloid and Interface Science*, vol. 111, p. 378, 1986.

[VAN 88] VAN OSS C.J., CHAUDHURY M.K., GOOD R.J., "Interfacial Lifshitz-Vanderwaals and polar interactions in macroscopic systems", *Chemical Reviews*, vol. 88, p. 927, 1988.

[VER 48] VERWEY E.J., OVERBEEK J.T., *Theory of Stability of Lyophobic Colloids*, Elsevier, Amsterdam, 1948.

[VER 99] VERHEIJEN H.J.J., PRINS M.W.J., "Reversible electrowetting and trapping of charge: model and experiments", *Langmuir*, vol. 15, p. 6616, 1999.

[VIT 00] VITTOZ C., DUBOIS P.E., JOUD J.C., "Third international conference on high temperature capillarity Kurashiki, Japan", Conference Invitée, *Proceeding of the 3rd Conference on High Temperature Capillarity*, pp. 50–62, 19–22 November 2000.

[VIT 96] VITTOZ C., Etude de l'acido-basicité des surfaces par mouillabilité et XPS: Applications aux aciers inoxydables, PhD Thesis, INP Grenoble, 1996.

[WAT 91] WATTS J.F., GIBSON E.M., "The determination of the acid-base properties of inorganic surfaces by X-ray photoelectron spectroscopy", *International Journal of Adhesion and Adhesives*, vol. 11, pp. 105–108, 1991.

[WEN 49] WENZEL R.N., *Industrial and Engineering Chemistry Research*, vol. 28, p. 988, 1936; *Journal of Physical and Colloid Chemistry*, vol. 53, p. 1466, 1949.

[WHY 08] WHYMAN G., BORMASHENKO E., STEIN T., "The rigourous derivation of Young, Cassie-Baxter and Wenzel equations and the analysis of the contact angle hysteresis phenomenon", *Chemical Physics Letters*, vol. 450, pp. 355–359, 2008.

[WU 82] WU S., *Polymer Interface and Adhesion*, M. Dekker, Inc, New York, 1982.

[YOU 05] YOUNG T., "An essay on the cohesion of fluids", *Philosophical Transactions of the Royal Society of London*, vol. 95, p. 65, 1805.

Index

A, B, C

D, E, F

G, H, I

J, K, L

M, N, O

P, R, S

T, V

W, Y, Z

Other titles from

ISTE

in

Materials Science

2015

LALEVÉE Jacques, FOUASSIER Jean-Pierre
Dyes and Chromophores in Polymer Science

2014

HAMAIDE Thierry, DETERRE Rémi, FELLER Jean-François
Environmental Impact of Polymers

2013

BATHIAS Claude
Fatigue Limit in Metals

LEXCELLENT Christian
Shape-memory Alloys Handbook

2012

DAVIM J. Paulo
Wear of Advanced Materials

DELHAES Pierre
Carbon Science and Technology: From Energy to Materials

GOUDEAU Philippe, GUINEBRETIÈRE René
X-Rays and Materials

SAANOUNI Khemais
Damage Mechanics in Metal Forming: Advanced Modeling and Numerical Simulation

VANBÉSIEN Olivier
Artificial Materials

VERDU Jacques
Oxidative Ageing of Polymers

2011

BATHIAS Claude, PINEAU André
Fatigue of Materials and Structures: Application to Damage

BATHIAS Claude, PINEAU André
Fatigue of Materials and Structures: Application to Design

PRIESTER Luisette
Grain Boundaries and Crystalline Plasticity

VIGNES Alain
Extractive Metallurgy 1: Basic Thermodynamics and Kinetics

VIGNES Alain
Extractive Metallurgy 2: Metallurgical Reaction Processes

VIGNES Alain
Extractive Metallurgy 3: Processing Operatiuons and Routes

2010

BATHIAS Claude, PINEAU André
Fatigue of Materials and Structures V1 / fundamental concepts

CHATEIGNER Daniel
Combined Analysis

CHEVALIER Yvon, VINH TUONG Jean
Mechanical Characterization of Materials and Wave Dispersion V2

CHEVALIER Yvon, VINH TUONG Jean
Mechanical Characterization of Materials and Wave Dispersion

DELHAES Pierre
Carbon-based Solids and Materials

2009

ALVAREZ-ARMAS Iris, DEGALLAIX-MOREUIL Suzanne
Duplex Stainless Steels

DAVIM J. Paulo
Machining Composite Materials

GALERIE Alain
Vapor Surface Treatments

2008

BATTAGLIA Jean-Luc
Heat Transfer in Materials Forming Processes

BLONDEAU Régis
Metallurgy and Mechanics of Welding

FRANÇOIS Dominique
Structural Components

REYNE Maurice
Plastic Forming Processes

TAKADOUM Jamal
Materials and Surface Engineering in Tribology

2007

BOCH Philippe, NIEPCE Jean-Claude
Ceramic Materials

CRISTESCU Constantin
Materials with Rheological Properties

2006

LASSEN Tom, RÉCHO Naman
Fatigue Life Analyses of Welded Structures

CPSIA information can be obtained
at www.ICGtesting.com
Printed in the USA
BVHW062242220122
626943BV00002B/67

9 781848 218437